THE WESTERN

PRACTICAL ARITHMETIC,

WHEREIN THE

RULES ARE ILLUSTRATED, AND THEIR PRINCIPLES

EXPLAINED:

CONTAINING

A GREAT VARIETY OF EXERCISES,

PARTICULARLY ADAPTED TO THE

CURRENCY OF THE UNITED STATES.

WITH

AN APPENDIX:

CONTAINING THE CANCELING SYSTEM, ABBREVIATIONS IN MULTIPLICATION, MENSURATION, AND THE ROOTS.

DESIGNED FOR THE USE OF

SCHOOLS AND PRIVATE STUDENTS.

COMPILED

BY JOHN L. TALBOTT.

CINCINNATI:
PUBLISHED BY E. MORGAN & CO.
Stereotyped by J A. James.
1851.
Price 37½ Cents.

PREFACE.

In presenting this work to the public, the author makes no pretensions to having discovered any new spring by which to put the youthful mind into action, nor any new method of communicating a knowledge of Arithmetic. He has founded his work on the belief that *labor* and *labor only*, can insure success in any pursuit; and that labor should always be bestowed upon those objects which produce the greatest useful result.

In the selection and arrangement of matter, therefore, those rules that are ot the most general use, have been presented first, and their exercises made extensive, that the pupil many early become familiar with their principles, and expert in their application.

The explanations accompanying the rules, are designed to facilitate the progress of private students, and to diminish the labor of teachers, especially in large schools, where they are unable to give to each pupil the necessary explanations.

The Mensuration of *Carpenters'*, *Masons'*, *Plasterers'* and *Pavers'* work, &c., will be found an acceptable part of Arithmetic, to every man of business, and a practical knowledge of it will contribute much to the security and satisfaction of both workmen and employers, in estimating amounts of work. This has been introduced in consequence of numerous applications to the author to measure various kinds of work, and for instruction in particular rules of Mensuration.

The system of *Book Keeping*, is thought to be sufficient for all the purposes of *farmers*, *mechanics* and *retailers*, in that necessary branch of a business education.

How far the author has succeeded in his attempts to compile a *useful work*, particularly adapted to the circumstances of the Western People, remains for *them* to judge, and for *experience* to determine.

NOTICE.

The favorable reception of this treatise and the increasing demand for it, have induced the publishers to revise, enlarge, and otherwise improve the work. Such alterations and amendments have been made as the experience of the author and of other intelligent and successful teachers has suggested; it is therefore presumed, that the work will be found more useful, and consequently more acceptable than heretofore.

Numerous testimonials to the merits of the work, have been received; but its general adoption without any efforts to force its introduction, and its intrinsic worth, are our main reliance; we have therefore given it a thorough revision, and now submit the result of our labors to a discerning public.

February, 1841. F MORGAN & Co.

CONTENTS.

ARITHMETIC.

Arithmetic is that part of Mathematics which treats of numbers. It is both a *science* and an *art*;—the *science* explains the nature of numbers, and the principles upon which the rules are founded, while the *art* relates merely to the application of the various rules.

All the operations of arithmetic are conducted by means of Five fundamental rules, viz., *Numeration*, (which includes *Notation*,) *Addition*, *Subtraction*, *Multiplication*, and *Division*.

NUMERATION AND NOTATION.

Numeration is the art of representing figures or numbers by words; Notation is the art of representing numbers by characters called figures.

All numbers are represented by the following characters, which are called *figures* or *digits*.

0,	1,	2,	3,	4,	5,	6,	7,	8,	9.
nought,	one,	two,	three,	four,	five,	six,	seven,	eight,	nine.

The *one* is often called a *unit*, it signifies a whole thing of a kind; two signifies two units or ones; three signifies three units or ones, &c.

The value which the figures have when standing alone is called their *simple* value; but in order to denote numbers higher than 9, it is necessary to give them another value called a *local* value, which depends entirely on the order or place in which they stand. Thus, when we wish to write the number *ten* in figures, we do it by combining the characters already known, placing a 1 on the left hand of the 0, thus, 10, which is read *ten*. This 10 expresses ten of the units denoted by 1, but as it is only a *single ten* it is called a *unit*, and the 1 being written in the second order or second place from the right hand to express it, it is called a unit of the *second order*, the first place being called the *place of units*, and the second, the *place of tens ; ten units of the first order making one unit of the second order.*

When units simply are named, units of the first order are always meant, when units of any other order are intended, the name of the order is always added.

Two tens or twenty,	are	written	20.
Three tens or thirty,	"	"	30.
Four tens or forty,	"	"	40.
Five tens or fifty,	"	"	50.
Six tens or sixty,	"	"	60.
Seven tens or seventy,	"	"	70.
Eight tens or eighty,	"	"	80.
Nine tens or ninety,	"	"	90.
Ten tens or one hundred,	"	"	100.

The numbers between 10 and 20, between 20 and 30, between 30 and 40, &c. may easily be expressed by considering the tens and units of which they are composed. Thus, *eleven* being composed of one ten and one unit, is expressed thus, 11, *twenty-three* being composed of two tens, and three units, is expressed thus, 23. &c.

Sixteen being 1 ten and 6 units, is written thus, 16.
Thirty-nine being 3 tens and 9 units, is written 39.
Sixty-four being 6 tens and 4 units, is written 64.
Ninety-five being 9 tens and 5 units, is written 95.

Ten tens or *one hundred* forms a unit of the third order; it is expressed by placing a 1 in the third place, and filling the first and second places with cyphers, thus, 100. Two hundred is expressed thus, 200 Three hundred thus, 300, &c.

With the orders of units, tens, and hundreds, all the numbers between one and one thousand may be readily expressed. For example, in the number four hundred and twenty-seven, there are 4 hundreds, 2 tens, and 7 units, that is, 4 units of the third order, 2 units of the second order, and 7 units of the first order.

	hun.	tens.	units.
Hence the number is written thus,	4	2	7

In the number three hundred and five, there are 3 hundreds, no tens, and 5 units, or 3 units of the third, none of the second, and five of the first order, hence the number is written thus,

hun.	tens.	units.
3	0	5

Ten units of the order of hundreds, that is ten hundreds form a unit of the fourth order, called *thousands*, written thus, 1000.

In the same manner ten units of the fourth order form a unit of the fifth order, called *tens of thousands.*

The following may be regarded as the principles of Notation and Numeration.

1st. *Ten units of the first or lowest order, make one unit of the second order; ten units of the second order, make one unit of the third order, and universally ten units of any order make a unit of the next higher order.*

2d. *All numbers are expressed by the nine digits, and the cypher, and this is effected by giving to the same figure different values according to the place it occupies.* Thus, 4 in the first place is 4 units, in the second place 4 tens or forty, and so on. This tenfold increase of value by changing the place of the same figure is usually expressed by saying that figures increase from right to left in a tenfold proportion. The names of the orders are to be learned from the

NUMERATION TABLE.

Hundreds of millions.	Tens of millions.	Millions	Hundreds of thousands.	Tens of thousands.	Thousands.	Hundreds.	Tens.	Units.
9	8	7	6	5	4	3	2	1

The orders are likewise divided into periods of six places each, according to the following table.

of Billions.						of Millions.						of Units.					
Hundreds of thousands.	Tens of thousands.	Thousands.	Hundreds.	Tens.	Units.	Hundreds of thousands.	Tens of thousands.	Thousands.	Hundreds.	Tens.	Units.	Hundreds of thousands.	Tens of thousands.	Thousands.	Hundred.	Tens.	Units.

The periods succeeding those in the table, are *Trillions*, *Quadrillions*, *Quintillions*, *Sextillions*, *Septillions*, *Octillions*, and *Nonnillions*, and analogical names might be formed for the succeeding higher periods.

From the preceding remarks the pupil will readily understand the reason of the following rule for numerating or expressing figures by words.

Rule.—Commence at the right hand, and separate the given number into periods, then beginning at the left hand, read the figures of each period as if they stood alone, and then add the name of the period.

Thus, the number 8304000508245, when divided into periods, becomes 8,304000,508245, and is read, *Eight billion, three hundred and four thousand million, five hundred and eight thousand two hundred and forty-five.* The name *unit* of the right hand period is commonly omitted in reading.

EXERCISES IN NUMERATION.

Ex. 1.	35	10.	3700054	19.	20031025
2.	204	11.	6130425	20.	68723145
3.	513	12.	2701030	21.	901023406
4.	2000	13.	3705423	22.	820302008
5,	3054	14.	6803217	23.	310275603
6.	7428	15.	2003005	24.	600000501
7	10345	16.	70032004	25.	3000400230024
8.	40024	17.	62003005	26.	80000102051003
9.	61304	18.	91010010	27.	50000021375604

28. 4000012000040250014

29. 1000982000375000482000354000271000032561804

From the preceding tables and remarks, the pupil will likewise readily understand the reason of the following rule for notation, or expressing numbers by figures.

RULE.—Make a sufficient number of cyphers or dots, and divide them into periods, then underneath these dots write each figure in its proper order and fill the vacant orders with cyphers.

NOTE.—The object of the dots or cyphers, being to guide the learner at first, after a little practice he may dispense with them.

Ex. 1. Write down in figures the number twenty millions three hundred and four thousand and forty. Here millions being the highest period named, we write cyphers to correspond with that, and the period of units, and then underneath these place the significant figures in their proper order, and afterwards fill the vacant orders with cyphers.

0 0 0 0 0 0, 0 0 0 0 0 0
2 0 3 0 4 0 4 0

The pupil must recollect that cyphers being of no use except to fill vacant orders, are never to be placed to the left of whole numbers.

EXERCISES IN NOTATION.

Express the following numbers in figures.

EXAMPLES.

2. Seventy-five.
3. Ninety.
4. One hundred and five.
5. Three hundred and twenty.
6. Nine hundred and four.
7. Eight hundred and ninety.
8. Two thousand three hundred and five.
9. Six thousand and forty.
10. Seven thousand and four.
11. Eight thousand and ninety-five.
12. Ten thousand five hundred and fifty-six
13. Forty thousand and forty.
14. Ninety-five thousand two hundred and sixty-seven.
15. Eighty thousand one hundred and nine.

16. One hundred and thirty-six thousand two hundred and seventy five.
17. Three hundred and seven thousand and sixty-four.
18. Five hundred thousand and five.
19. One million, two hundred and forty-seven thousand, four hundred and twenty-three.
20. Ten millions, forty thousand and twenty.
21. Sixty millions, seventeen thousand and two.
22. One hundred and four millions two hundred and four thousand and sixty-five.
23. Five hundred and three millions, one hundred and two thousand and nine.
24. Ninety one thousand and two millions, and four.
25. Sixty billions, three millions and forty-one thousand.
26. One billion, one hundred million, one thousand and one.

The *Roman* method of representing numbers, is by means of certain capital letters of the Roman alphabet. Thus:

I	one	XVIII	eighteen
II	two	XIX	nineteen
III	three	XX	twenty
IV	four	XXX	thirty
V	five	XL	forty
VI	six	L	fifty
VII	seven	LX	sixty
VIII	eight	LXX	seventy
IX	nine	LXXX	eighty
X	ten	XC	ninety
XI	eleven	C	one hundred
XII	twelve	CC	two hundred
XIII	thirteen	CCC	three hundred
XIV	fourteen	CCCC	four hundred
XV	fifteen	D	five hundred
XVI	sixteen	M	one thousand
XVII	seventeen	MDCCCXXXVIII	1838

NOTE 1. As often as any letter is repeated, so often is its value repeated.
NOTE 2. A less character before a greater one, diminishes its value
NOTE 3. A less character after a greater one, increases its value.

QUESTIONS.

What is Arithmetic? When is it a science? When is it an art? What are the fundamental rules of arithmetic? What is numeration? What is notation? What does a unit signify? What does two signify? Three, &c.? What is meant by the simple value of a unit? What does the local value of a figure depend on? How do you write the number ten in figures? Why is the one in this case called a unit of the second order? How many units of the first order does it take to make a unit of the second order? How many units of the second order does it require to form a unit of the third order? &c. Repeat the principles of notation and numeration. Repeat the names of each of the first nine orders as expressed in the numeration table. Repeat the name of each of the periods. Repeat the Rule for numeration. Repeat the Rule for notation.

EXPLANATION OF CHARACTERS.

Signs.	*Significations.*
$=$	equal; as 20s. $=$ £1.
$+$	more; as $6 + 2 = 8$.
$-$	less; as $8 - 2 = 6$.
$\times$	into, with, or multiplied by; as $6 \times 2 = 12$.
$\div$	by (*i. e.* divided by;) as $6 \div 2 = 3$; or, 2)6(3.
: :: :	proportionality; as 2 : 4 :: 6 : 12.
$\surd$ or, $\surd$	Square Root; as $\surd 64 = 8$.
$\surd$	Cube Root; as $\surd 64 = 4$.
$\surd$	Fourth Root; as $\surd 16 = 2$, &c.
———	A vinculum; denoting the several quantities over which it is drawn, to be considered jointly as a simple quantity.

SIMPLE ADDITION.

Simple addition is the art of collecting several numbers, of the same name, into one sum.

RULE.

Place the numbers with *units* under *units*, *tens* under *tens*, &c. Begin the addition at the units, or right hand column, and add together all the figures in that column; then, if the amount be less than *ten*, set down the whole sum: but if greater than ten, see how many tens there are, and set down the number above the even tens, and carry *one* for each *ten* to the next column, and proceed with it as in the first.

Proof.—Begin the addition at the top of each column, and proceed as before, and if the result be the same, it is presumed to be right.

EXAMPLES.

(1)	(2)	(3)	(4)
4 3 2	2 3 1	2 1 4	4
2 1 3	4 1 3	1 2 1	2
1 2 1	1 2 1	3 1 2	5
2 1 3	1 3 2	3 2 1	3
9 7 9 sum	8 9 7 sum	9 6 8 sum	1 4 sum

(5)

```
  2 7 6 3 6
  7 9 8 9 2
  3 8 9 4 1
  6 7 8 3 2
  5 9 2 4 4
-----------
2 7 3 5 4 5
```

Here 4, 2, 1, 2 and 6 make 15. In fifteen there is *one ten* and five units. Set down the five units under the units column, and carry *one* for the ten to the next or *tens* column.

Then 1, 4, 3, 4, 9 and 3 make 24; in 24 there are two tens, and four over: set down the four under the column of *tens*, and carry *two* to the next or *hundreds* column &c., to the last, where the whole amount may be set down.

(6)	(7)	(8)
47386	99786	72752
29492	86937	37823
18583	27849	78794
89294	49878	23567
28887	72937	98372
74392	48732	12345
288034	386119	323653

(9)	(10)	(11)
47823	72683	84736
73714	95892	78928
27834	82783	27849
23925	94973	63782
67883	76892	28637
62734	43987	73862

(12)	(13)	(14)
73684	9376	7379
75	723	7463
473	87	729
6893	9	489
7	48	72
483	937	68
96	92	432

APPLICATION.

1. Add 224 dollars, 365 dollars, 427 dollars, and 784 dollars, together.

	224
	365
	427
	784
Answer,	$ 1800 Dollars

2. Add 3742 bushels, 493 bushels, 927 bushels, 643 bushels, and 953 bushels, together.

Answer, 6758 bushels.

3. Add 7346 acres, 9387 acres, 8756 acres, 8394 acres, 32724 acres. Ans. 66607 acres.

4. Henry received at one time 15 apples, at another 115, at another 19. How many did he receive?

Ans. 149.

5. A person raised in one year 724 bushels of corn, in another 3498 bushels, in another 9872. How much in all? Ans. 14094 bushels.

6. A man on a journey, travelled the first day 37 miles, the second 33 miles, the third 40 miles, the fourth 35 miles. How far did he travel in the four days?

Ans. 145 miles.

7. A has a flock of sheep containing 34. B has a flock of 47, and C of fifty-four. How many sheep are there in the three flocks? Ans. 135.

8. The distance from Philadelphia to Bristol is 20 miles; from Bristol to Trenton, 10 miles; from Trenton to Princeton, 12 miles; from Princeton to Brunswick, 18 miles; from Brunswick to New York, 30 miles. How many miles from Philadelphia to New York? Ans. 90.

9. A person bought of one merchant, 10 barrels of flour, of another 20 barrels, of another 95 barrels. How many barrels did he buy? Ans. 125 barrels.

10. A wine-merchant has in one cask 75 gallons, in another 65, in a third 57, in a fourth 83; in a fifth 74, and in a sixth 67 gallons. How many gallons has he in all? Ans. 421 gallons.

Questions.

How many primary rules of Arithmetic are there?

What are they called?

What is addition?

How do you place numbers to be added?

Where do you begin the addition?

Why do you carry one for *ten*, in preference to any other number?

Ans. Because it takes ten *ones* to make one *ten*, ten *tens* to make one *hundred*, &c. (*See table, page* 9.)

SIMPLE SUBTRACTION.

SIMPLE SUBTRACTION is taking a less number from a greater, of the same name, to show the difference between them.

The *greater* number is called the *minuend.*

The *less* number is called the *subtrahend.*

The *difference,* or what is left, is called the *remainder*

RULE.

Place the *less* number under the *greater*, with *units* under *units, tens* under *tens,* &c.

Then draw a line under them; begin at the right hand or units place, and subtract each figure of the subtrahend from the figure of the minuend that is above it, and set the remainder below. When the figure in the subtrahend is greater than the one above it, borrow one (which is one *ten*) from the next figure, and add it to the figure of the minuend; then subtract from the sum.

Proof.—Add the remainder and the subtrahend together, and if the sum equal the minuend, the work is presumed to be right.

EXAMPLES.

(1)		(2)
7 9 2 5 2 7 4 3	Minuend	9 7 3 8 4 7 6
3 4 1 2 0 3 1 2	Subtrahend	2 6 1 4 2 5 3
4 5 1 3 2 4 3 1	Remainder	7 1 2 4 2 2 3

(3)

7 2 6 3 9 8 2
6 4 2 5 6 3 7
8 3 8 3 4 5

Here we cannot take seven from two; then we must borrow one from the 8: that one is one *ten;* then ten and two are twelve; now take seven from twelve, and five remain.

One is borrowed from the 8, leaving only 7; then take 3 from 7, and 4 remain: or, suppose 8 to remain undiminished; and to cancel the one which is borrowed from the 8, add one to the 3 below, making four; then four from eight and four remain, as before, &c.

(4)	(5)
9 2 7 3 8 4 7	8 2 7 0 3 6 8 2
2 6 4 1 3 8 6	2 7 3 4 1 2 3 7
6 6 3 2 4 6 1	5 5 3 6 2 4 4 5

(6)	(7)
7 8 3 7 2 8 6	2 7 3 6 8 3 0 7 0
3 2 7 3 1 9 5	4 3 2 1 7 2 5
4 5 6 4 0 9 1	2 6 9 3 6 1 3 4 5

(8)	(9)
6 8 4 2 7 3 6 2	5 9 3 7 8 4 2 8 3
3 4 6 1 3 5 2 4	5 4 3 2 1 4 3 2

(10)	(11)
7 9 2 8 3 6 8 4 2	9 2 0 3 7 8 4 2
2 4 6 5 3 1 2 8	4 1 3 7 2 7 6 1

APPLICATION.

1. From 78 take 32 and what will remain?
Answer, 46.

2. From 478 take 324. What will remain?
Ans. 154.

3. Charles had 723 apples, and sold 421. How many has he left? Ans. 302

4. James had 9768 dollars, and gave for a house and lot 3453 dollars. How many has he left? Ans. 6315.

5. A farmer had 3849 acres of land; he gave to his sons 2135 acres. How many acres has he left for himself? Ans. 1714.

6. There are two piles of bricks, one contains 7896, and the other 4389. How many more are there in the one than in the other? Ans. 3507.

7. Bought 100 bags of coffee, weighing 14510 lbs., and sold thereof 63 bags weighing 6871 pounds; how many bags, and how many pounds remain unsold?

Ans. 37 bags, and 7639 lbs.

8. A man bought a chaise for 175 dollars, and to pay for it gave a wagon worth 37 dollars, and the rest in money. How much money did he pay?

Ans. 138 dollars.

9. A man deposited in bank 8752 dollars, and drew out at one time 4234 dollars, at another 1700 dollars, at another 962 dollars, and at another 49 dollars. How much had he remaining in bank? Ans. 1807 dollars.

10. A merchant bought 4875 bushels of wheat, and sold 2976 bushels. How many bushels remain in his possession? Ans. 1899.

11. A grocer bought 25 hogsheads of sugar, containing 250 hundred weight, and sold 9 hogsheads, containing 75 hundred weight. How many hogsheads and how many hundred weight had he left?

Ans. 16 hogsheads, and 175 hundred weight.

12. A traveller who was 1300 miles from home, travelled homeward 235 miles in one week; in the next 275 miles; in the next 325 miles; and in the next 290 miles. How far had he still to go, before he would reach home?

Ans. 175 miles.

Questions.

What is subtraction?

What is the greater number called?

What is the less number called?

What is the difference called?

How do you place numbers for subtraction?

Where do you begin the subtraction?

When the lower figure is greater than the upper one, how do you proceed?

Why is the *one* you borrow, one *ten*.

Ans. Because ten *ones* make one *ten;* and if I borrow one *ten* it will make ten *ones* again, &c.

How do you prove subtraction?

SIMPLE MULTIPLICATION.

SIMPLE MULTIPLICATION is a short method of performing particular cases of addition.

The number to be multiplied, is the *multiplicand.*

The number to be multiplied by, is the *multiplier.*

The number produced is the *product.*

The multiplicand and multiplier are sometimes called *factors.*

MULTIPLICATION TABLE.

Twice		3 times		4 times		5 times		6 times		7 times	
1 make	2	1 make	3	1 make	4	1 make	5	1 make	6	1 make	7
2	4	2	6	2	8	2	10	2	12	2	14
3	6	3	9	3	12	3	15	3	18	3	21
4	8	4	12	4	16	4	20	4	24	4	28
5	10	5	15	5	20	5	25	5	30	5	35
6	12	6	18	6	24	6	30	6	36	6	42
7	14	7	21	7	28	7	35	7	42	7	49
8	16	8	24	8	32	8	40	8	48	8	56
9	18	9	27	9	36	9	45	9	54	9	63
10	20	10	30	10	40	10	50	10	60	10	70
11	22	11	33	11	44	11	55	11	66	11	77
12	24	12	36	12	48	12	60	12	72	12	84

8 times		9 times		10 times		11 times		12 times	
1 make	8	1 make	9	1 make	10	1 make	11	1 make	12
2	16	2	18	2	20	2	22	2	24
3	24	3	27	3	30	3	33	3	36
4	32	4	36	4	40	4	44	4	48
5	40	5	45	5	50	5	55	5	60
6	48	6	54	6	60	6	66	6	72
7	56	7	63	7	70	7	77	7	84
8	64	8	72	8	80	8	88	8	96
9	72	9	81	9	90	9	99	9	108
10	80	10	90	10	100	10	110	10	120
11	88	11	99	11	110	11	121	11	132
12	96	12	108	12	120	12	132	12	144

CASE 1.

When the multiplier does not exceed 12.

RULE.—Place the *multiplier* under the *units* figure of the *multiplicand;* and multiply each figure of the multiplicand in succession, and set down the amount, and carry, as in addition.

Proof.—Multiply the *multiplier* by the *multiplicand.*

EXAMPLES.

4 2 3 1 Multiplicand	3 4 2 5 3	7 3 4 2
2 Multiplier	3	4
8 4 6 2 Product	1 0 2 7 5 0	2 0 3 6 8

3 6 5 6 3	8 3 7 5	4 3 7 8	9 2 8 6
5	6	7	8
1 8 2 8 1 5	5 0 2 5 0	3 0 6 4 6	7 4 2 8 8

4 3 7 5	7 8 6 2	3 7 2 4	7 4 8 2
9	1 0	1 1	1 2
3 9 3 7 5	7 8 6 2 0	4 0 9 6 4	8 9 7 8 4

EXERCISES

1 Multiply	4218	by	2	Product 8436
2	7321	by	3	21963
3	87692	by	4	350768
4	900078	by	9	8100702
5	826870	by	10	8268700
6	278976	by	11	3068736
7	569769	by	12	6837228

Case 2.

When the multiplier exceeds 12.

Rule.—Place the multiplier as before, with *units* under *units*, &c. Then multiply all the figures of the multiplicand by the *units* figure of the multiplier, setting down the product as before.

Proceed with the *tens* figure in the same manner, observing to set the product of the first figure in the tens place and with the hundreds figure placing the first product in hundreds place, &c., and add the several products together.

EXAMPLES.

```
  43752
    436
--------
 262512
131256
175008
--------
19075872
--------
```

Here we multiply by the 6 or *units* figure as before: then by the 3 or *tens* figure, placing the first product in the second or *tens* place, immediately under the three, in the multiplier. In like manner we use the 4, placing the first product in the third or *hundreds* place, immediately under the 4; after which we add the several products together, and the work is done.

```
   73684           37462
     427             563
--------        --------
  515788          112386
 147368          224772
294736          187310
--------        --------
31463068        21091106
--------        --------
```

EXERCISES.

1	Multiply	4736	by	34	Product 161024
2		5762	by	43	247766
3		6483	by	54	350082
4		7368	by	45	331560
5		4327	by	56	242312
6		7382	by	67	494594
7		4728	by	76	359328
8		7584	by	87	659808
9		5678	by	78	442884
10		7683	by	89	683787
11		4962	by	98	486276
12		7384	by	87	642408
13		4376	by	97	424472
14		7923	by	78	617994
15		6842	by	89	[illegible]38
16		7648	by	523	3999904
17		8473	by	456	3863688
18		9372	by	567	531392[illegible]

NOTE 1.—When either or both of the factors have noughts on the right hand, they may be omitted in the operation, and annexed to the product. Thus:

```
   47 | 000          734 | 00
   42 | 00            42 | 000
  ----------        ------------
   94               1468
  188              2936
  ----------       ------------
Product 197400000  Product 3082800000
```

NOTE 2.—When the multiplier is the exact product of any two factors in the multiplication table, the operation may be performed by separating the multiplier into its components, and multiplying first by the one, then its product by the other. Thus:

```
754 by 36     754       754      754
    9           3         6       36
-----        -----     -----    -----
 6786         2272      4524     4524
    4           12         6    2262
-----        -----     -----    -----
27144        27144     27144    27144
```

9 and 4, or 3 and 12, or 6 and 6, multiplied together, produce 36, and by using either pair, according to the above note, the true result is obtained.

EXERCISES.

1 Multiply	756	by	42	Product 31752
2	645	by	24	15480
3	876	by	48	42048
4	963	by	56	53928
5	827	by	72	59544
6	946	by	81	76626
7	875	by	84	73500
8	948	by	96	91008
9	795	by	108	85860

The pupil may work these by all the several pairs of components that he can find in the multiplier.

NOTE 3.—When the multiplier is not the exact pro duct of any two numbers in the table, use two factors whose product is short of the multiplier, then multiply the sum by the number required to supply the deficiency and add its product to that obtained by the two factors.

EXAMPLES.

583×3	583×2	583×5	583
4	7	3	23
2332	4081	1749	1749
5	3	6	1166
11660	12243	10494	13409
1749	1166	2915	
13409	13409	13409	

EXERCISES.

1 Multiply	846	by	26	Product 21996
2	784	by	29	22736
3	975	by	34	33150
4	859	by	43	36937
5	794	by	59	46846

PROMISCUOUS EXERCISES.

1. Charles has 24 marbles, and John has 13 times as many; how many has John? Ans. 312.

2. A gentleman owns 17 houses, for each of which he receives 250 dollars rent; how much does he receive for them all? Ans. 4250.

3. A laborer hired himself to a farmer for 11 years, at 150 dollars a year; how much did he receive? Ans. 1650 dollars.

4. A person wishes to purchase 26 shares of Bank stock at 75 dollars a share; what must he pay? Ans. 1950 dollars.

5. A mason having built a house, found that 98470 bricks were in it; suppose he desires to build 19 such houses, how many bricks must he obtain for the purpose? Ans. 1870930.

SIMPLE DIVISION.

Simple division is a short method of performing several subtractions.

The number to be divided is called the *dividend.*

The number by which it is to be divided is called the *divisor.*

The number of times that the divisor is contained in the dividend, is called the *quotient.*

So many figures of the dividend as are taken to be divided at one time, is called a *dividual.*

If any thing remain when the operation is completed, it is called the *remainder.*

Case i.—Short Division.

When the divisor does not exceed 12.

Rule.—Place the *divisor* on the left hand side of the number to be divided.

Consider how often the divisor is contained in the first figure or figures of the *dividend,* and set down the result below; observing how many remain, if any. If there be no remainder, consider how often the divisor is contained in the next figure: but if there be a remainder, call it so many *tens,* and add the next figure to it, and divide the sum, placing the result beneath, as before.

Proof.—Multiply the quotient by the divisor; add the remainder, if any, and the product will equal the dividend.

EXAMPLES.

	Dividend.			
Divisor	2)182	2)648	3)963	4)484
Quotient	241	324		

```
      3)741851
        ------
        247283+2 Remainder
             3
        ------
Proof   741851
```

Here 3 are contained in 7 *two times*, and one remains; place the *two* under the *seven*, and suppose the *one* that remains to be one *ten*, and add the next figure (4) to it, which makes *fourteen.*

Now 3 are contained in fourteen 4 times, and 2 remain. Set the 4 down under the 4 in the dividend, and suppose the *two* that remain to be two *tens*, and add the next figure (1) to it, which make *twenty-one.* Now 3 into 21 go 7 times, and no remainder. Place the 7 under the 1 in the dividend, and proceed in the same manner with the other figures.

```
      4)65270167                  5)6572686
        --------                    -------
        16317541+3 Rem.             1314537+1 Rem
               4                          5
        --------                    -------
Proof   65270167            Proof   6572686


      6)8739627                   7)4873692
        -------                     -------

        -------                     -------


      8)9273684                   9)8379286
        -------                     -------

        -------                     -------


10)946873          11)893726          12)98796
   ------             ------             -----

   ------             ------             -----
```

EXERCISES.

Divide	7893762	by	6	Ans.	1315627
	9387984	by	7		1341140—4
	6928437	by	8		866054—5
	9276874	by	9		1030763—7
	8672934	by	10		867293—4
	6873842	by	11		624894—8
	7369287	by	12		614107—3

Case 2.—Long Division.

When the divisor exceeds 12.

Rule.—Place the divisor to the left hand of the dividend, as in case 1.

Consider how often the divisor is contained in the least number of figures into which it can be divided; and set down the result at the right hand of the dividend.

Multiply the divisor by the quotient figure thus found, and set the product under the dividual or figures supposed to be divided.

Subtract the product from the dividual, and set down what remains. Bring down the next figure of the dividend, and proceed as before, till all the figures are brought down and divided.

EXAMPLES.

```
Divisor. Dividend. Quotient.
   27)984376(36458
      81
      ——
      174
      162
      ——
       123
       108
       ——
        157
        135
        ———
         226
         216
         ——10 Remainder.
```

Twenty-seven into 98 go 3 times: multiply the divisor (27) by 3 and set the product under the dividual (98) and subtract. To the remainder (17) bring down the next figure (4) of the dividend. Now 27 into 174 go 6 times. Place the 6 in the quotient and multiply (27) the divisor, by 6, and set the product under 174 and subtract as before, &c.

```
Divisor. Dividend. Quotient.
  42)    98754    (2351
          84          42 Divisor
          ---       ----
          147       4702
          126      9404
          ---         12 Remainder
           215     -----
           210     98754 Proof
           ------------
             54
             42
             --
             12 Remainder

32)789627(24675            65)1827538(28115
   64                         130
   ---                        ---
   149                         527
   128                         520
   ---                         ---
    216                         75
    192                         65
    ---                         ---
     242                        103
     224                         65
     ---                        ---
      187                       388
      160                       325
      ---                       ---
       27 Rem.                   63 Rem.
```

EXERCISES.

Divide	8769	by	13	Quo. 674	Rem. 7
	476	by	15	31	11
	958	by	18	53	4
	1475	by	28	52	19
	4277	by	31	137	30
	25757	by	37	696	5
	63125	by	123	513	26
	253622	by	422	601	

NOTE 1.—Cyphers on the right hand of the divisor may be omitted in the operation, observing to separate as many figures from the right of the dividend, which must be annexed to the remainder.

EXAMPLES.

```
54 | 00)1463 | 40(27          32 | 0)7617 | 3(238
      108                          64
      ---                          --
      383                          121
      378                           96
      ---                          ---
  Rem. 540                         257
                                   256
                                   ---
                               Rem. 13
```

EXERCISES.

				Ans.	Rem.
Divide	40220	by	1900	21	320
	137000	by	1600	85	1000
	99607765	by	27000	3689	4765
	2304108	by	5800	397	1508

NOTE 2.—When the divisor is the exact product of any two numbers in the multiplication table, the operation may be performed by dividing first by one of the component parts, and then the quotient by the other.

To get the true remainder, multiply the last remainder by the first divisor, and add the first remainder.

EXAMPLES.

```
   (7 | 98754
42 {   -----
   (6 | 14107—5 first remainder
        -----
        2351—1 last remainder
             7 first divisor
             -
      add {  7
          {  5 first remainder
             -
            12 true remainder
```

$$27 \begin{cases} 3 \\ 9 \end{cases} \begin{array}{|l} 984376 \\ \hline 328125—1 \\ \hline 36458—3 \times 3 + 1 = 10 \text{ Rem} \end{array}$$

EXERCISES.

Divide				Quotient	Rem.
	9756	by	35	278	26
	8491	by	81	104	67
	44767	by	18	2487	1
	92017	by	56	1643	9
	38751	by	48	807	15
	734071	by	72	10195	31

APPLICATION.

1. Seven boys have 161 apples, which they divide equally among them. How many does each have?
Answer, 23.

2. What is the quotient, if 8736 be divided by 8, and that quotient by 4? Ans. 273.

3. If 350 dollars be equally divided among 7 men, what will be the share of each? Ans. 50.

4. How many times are 27 contained in 952?
Ans. 35 times and 7 over.

5. Suppose 2072 trees planted in 14 rows. How many trees will there be in each row? Ans. 148.

6. Several boys who went to gather nuts, collected 4741, of which each boy received 431. How many boys were there? Ans. 11.

7. If the expense of erecting a bridge, which is 15036 dollars, be equally defrayed by 179 persons, what must each pay? Ans. 84 dollars.

8. Suppose a man receive in one year 2920 dollars; how much a day is his income at that rate; and suppose that his expenses for the year amount to 1769 dollars. How much will he save in a year?

Ans. His income will be 8 dollars a day; he will save 1151 dollars in a year.

Questions.

What is division?

What do you call the number that is to be divided?

What do you call the number you divide by?

What do you call the number obtained by division?

What do you call that which is left when the work is done?

When the divisor does not exceed 12, how do you perform the operation?

When the divisor exceed 12, how do you proceed?

How do you prove division?

How may the operation be performed when there are cyphers at the right hand of the divisor?

How may it be performed when the divisor is the exact product of two numbers in the multiplication table?

How do you obtain the true remainder in the last case?

PROMISCUOUS EXERCISES IN THE PRECEDING RULES.

1. If the contents of five bags of dollars, containing $295, $410, $371, $355, and $520, be divided equally among 25 persons, how much is the share of each? Ans. $78.04

2. A man possessed of an estate of $30,000, disposed of it in the following manner: to his brother he gave $1500, and the balance to his 5 sons, to be equally divided among them. What was each one's share? Ans. $5700.

3. What number is it, which being added to 9709, will make 110901? Ans. 101192.

4. Add up twice 397, three times 794, four times 31196, five times 15880, six times 95280, and once 33304. Ans. 812,344.

5. Three merchants have a stock of 14876 dollars, of which A owns 4963 dollars, B 5188, and C the remainder. How much does C own? Ans. 4725 dolls.

FEDERAL MONEY,

OR MONEY OF THE UNITED STATES.

TABLE.

10 mills make	1 cent
10 cents	1 dime
10 dimes	1 dollar
10 dollars	1 eagle

These denominations bear the same relation to each other as those of units, tens, hundreds, &c. Federal money is therefore added, subtracted, multiplied, and divided by the same rules as Simple Addition, Subtraction, Multiplication, and Division.

ADDITION OF FEDERAL MONEY.

Rule.

Place the numbers one under another, with mills on the right, cents, dimes, &c., in succession; observing to keep *mills* under *mills*, *cents* under *cents*, &c. Then proceed as in simple addition.

When halves or fourths of a cent occur, find their amount in fourths, and consider how many cents these fourths will make, and carry them to the column of cents.

EXAMPLES.

Eagles.	*Dolls.*	*Dimes.*	*Cents*	*Mills.*		*Dolls.*	*Ds.*	*Cts.*
3	7	8	9	5		7	8	9
7	4	9	8	7		9	7	8
2	3	8	7	9		6	8	4
8	8	9	8	6		6	3	7
4	7	8	9	8		4	8	2
					E.			
27	3	6	4	5	3	5	7	0

NOTE.—In common business transactions, eagles, dimes, and mills are not used: dollars, cents, and fractions of a cent, are the only denominations kept in accounts.

EXAMPLES.

Ds.	cts.		Ds.	cts.
34	62		427	68
56	31		342	31
27	82		427	26
23	68		793	84
27	42		273	42
169	85		2264	51

EXERCISES.

(1) Ds.	cts.	(2) Ds.	cts.	(3) Ds.	cts.
468	31	927	24	273	45
723	62	768	32	846	37
845	92	427	56	283	75
736	25	792	34	846	91
846	31	587	62	674	75
428	62	842	27	273	25

Ds.	cts.
437	62½
386	81¼
243	18¾
427	37½
428	12½
1923	12½

One half is two-fourths; and one half more make four fourths, and three fourths more make seven fourths, and one fourth more make eight fourths, and one half (or two fourths) more make *ten fourths.* Four fourths make one cent, then ten fourths make two cents, and leave two fourths, or one half cent. Set down the ½ cent, and carry the two cents to the next column.

Ds.	cts.	Ds.	cts.	Ds.	cts.
274	81¼	27	68¾	56	06¼
362	87½	36	81¼	32	12½
421	18¾	28	62½	36	25
625	31¼	37	93¼	42	62½
241	56¾	24	62½	54	81¼

APPLICATION.

1. Add 48 dollars 20 cents; 14 dollars 58 cents; 100 dollars 25 cents; and 84 dollars 36 cents.

Ans. 247 dollars 39 cents.

2. Add $7,62½, $34,31¼, $72,06¼, $41,31¼, $25,68¾, and $87,43¾ together, and tell the amount.

Ans. $268,43¾.

3. Bought a hat for $4,25 cents; a pair of shoes for $2,25; a pair of stockings for $1,25, and a pair of gloves for 75 cents. What is the cost of the whole?

Ans. $8 50 cents.

4. If I buy coffee for $1,18¾, tea for $2,50, cloves for 87½, mace for 93¾, cinnamon for $1,87½, raisins for $2,68¾, nutmegs for 37½, candles for 87½, and wine for $1,93¾, what must I pay for them? Ans. $13,25.

Questions.

What relation do mills, cents, dimes, &c., bear to each other?

How are the addition, subtraction, multiplication, and division of Federal money performed?

How do you place the numbers to be added?

How do you proceed when halves, fourths, &c., occur?

SUBTRACTION OF FEDERAL MONEY.

Rule.—Place the less under the greater, with *dollars* under *dollars*, and *cents* under *cents;* then, if there are no fractions, proceed as in simple subtraction.

If there is a fraction in the upper sum and none in the lower, set it down as a part of the remainder, and proceed as before.

If there is a fraction in each sum, and the lower be less than the upper, subtract the lower from the upper, and set down the difference.

If the lower fraction be greater than the upper one, borrow one cent, and call it four fourths, and add them to the upper fraction, and subtract the lower one from the sum.

Proof.—As in simple subtraction.

EXAMPLES.

Ds.	*cts.*	*Ds.*	*cts.*	*Ds.*	*cts.*
32	62	43	68¾	75	68¾
21	31	21	25	24	12½
$11	31	$22	43¾	51	56¼

Ds.	*cts.*
271	62½
132	93¾
138	68¾

NOTE.—*Three* fourths cannot be taken from *two* fourths: then borrow one *cent* from the two cents, which has four fourths in it: add the *four* fourths to the *two* fourths, this makes *six* fourths; subtract *three* fourths from *six* fourths, and three fourths (¾) remain. Set down the ¾ and add one to the next 3, as in simple subtraction.

EXERCISES.

Ds.	*cts.*	*Ds.*	*cts.*	*Ds.*	*cts.*
65	49	520	31¼	436	31¼
35	12½	210	12½	243	18¾

Ds.	*cts.*	*Ds.*	*cts.*	*Ds.*	*cts.*
273	62½	237	56¼	732	31¼
124	37½	142	87½	261	68¾

APPLICATION.

1. Subtract $432,68¾ from 1000,93¾.
Ans. $568,25.

2. Subtraction shows the difference between two numbers; what is the difference between $37,62½ and $93,87½.
Ans. $56,25.

3. Bought goods to the amount of $545,95, and paid at the time of purchase $350. How much remains unpaid?
Ans. $195,95.

4. A merchant bought a quantity of coffee, for which he paid $560. He afterwards sold it for $610,87½ How much did he gain by the transaction?
Ans. $50,87½.

Questions.

How do you place the numbers in subtraction of Federal Money?

How do you perform the operation?

If a fraction occur in the upper line or minuend, what do you do with it?

If a fraction occur in each, how do you proceed?

Suppose the lower fraction is greater than the upper one, how do you proceed?

How do you prove subtraction of Federal Money?

MULTIPLICATION OF FEDERAL MONEY.

Rule.—Set the multiplier under the multiplicand, and if there be no fractions, proceed as in simple multiplication; observing to separate the cents from the dollars in the product.

If there is a fraction in the sum, multiply it, and see how many cents are in the product; set down the fraction that is over, and proceed as before.

Or if the multiplier exceeds 12, multiply the sum, omitting the fractions; then multiply the fraction, and add the number of cents contained in the product, to the product of the rest of the sum.

EXAMPLES.

Ds. cts.	*Ds. cts.*	*Ds. cts.*
12 , 50	10 , 56¼	23 , 62½
4	2	5
\$50 , 00	\$21 , 12½	\$118 , 12½

Ds. cts.	
10,87½	125 times
125	one half make
	125 halves: 2
5435	into 125 go 62
2174	times, leaving
1087	one; that is,
62½	one half, mak-
	ing 62½ cents.
\$1359,37½	

Ds. cts.	
4,18¾	24 times ¾ are
24	72 fourths: four
	fourths are con-
1672	tained 18 times in
836	72 fourths, mak-
18	ing 18 cents.
\$100,50	

EXERCISES.

1 Multiply	\$145,18¾	by	7	Ans. \$1016,31¼
2	7,87½	by	47	370,12½
3	28,68¾	by	68	1950,75
4	42,31¼	by	58	2454,12½
5	137,62½	by	67	9220,87½
6	79,00½	by	207	16354,03½

APPLICATION.

1. What will 8 pounds of cheese come to, at 18 cents a pound? Ans. \$1 44 cts.

2. What is the value of 12 yards of linen, at 35 cents a yard? Ans. \$4 20 cts.

3. What cost 29 yards of cloth at \$2 25 cts. a yard? Ans. \$65 25 cts.

4. What will 213 barrels of flour cost, at \$5 25 cents a barrel? Ans. \$1118 25 cts.

5. Bought 321 barrels of cider at \$1 25 cts. a barrel. What did it amount to? Ans. \$401 25 cts.

6. What will 580 bushels of salt cost at \$1 12½ cts. a bushel. Ans. \$652 50 cts.

7. What is the value of 2 pieces of cloth, one containing 38 yards, and the other 26 yards, at \$3 87½ cts. a yard? Ans. \$248.

8. What will be the cost of 132 pieces of linen at \$17 37½ cts. each? Ans. \$2293 50 cts.

9. What will 8 cords of wood amount to, at 4 dollars 50 cents a cord? Ans. 36 dollars.

10. Sold 213 barrels of flour for 6 dollars 25 cents per barrel. What is the amount? Ans. 1331 dols. 25 cts.

11. Bought 308 pounds of coffee at 21 cents a pound. What is the amount? Ans. 64 dols. 68 cts.

12. Bought 217 gallons of brandy at \$1 18¾ cts. per gallon; and sold it for \$1 37½ cts. per gallon. What was the amount paid for the whole; the sum it sold for; and the gain?

Ans. Prime cost, \$257 68¾: sold for \$298 37½; gain, \$40,68¾

DIVISION.

Rule.—Divide as in simple division. When a remainder occurs, multiply it by 4; and add the number of fourths that are in the fraction of the sum (if any) to its product: divide this product by the divisor, and its quotient will be fourths, which annex to the quotient.

Proof.—As in simple division.

EXAMPLES.

Ds. cts.	*Ds. cts.*	*Ds. cts.*
2)45,22	3)63,18¾	2)25,37½
22,61	21,06¼	12,68¾

Ds. cts.

```
25)629,68¾(25,18¾
   50
   ---
   129
   125
   ---
     46
     25
     ---
     218
     200
     ---
      18
       4
      --
      75
  25)75(3
```

Here 18 cents remain; multiply 18 cents by four, brings them to fourths of a cent; add the ¾, this makes 75 fourths: divide 75 fourths by 25, and ¾ are obtained, which place in the quotient.

```
32)78800(24,62½
   64
   ---
   148
   128
   ---
    200
    192
    ---
      80
      64
      --
      16
       4
      --
32)64(2 or ½
   64
```

EXERCISES.

	D. cts.			
Divide	56,15	by	10	Quotient $ 5,61½
——	96,00	by	5	—— 19,20
——	156,00	by	4	—— 39,00
——	58,14	by	38	—— 1,53
——	417,96	by	129	—— 3,24
——	494,45	by	341	—— 1,45
——	627,38	by	508	—— 1,23½

APPLICATION.

1. If 7 pounds of butter cost $1,89 cts., what is the value of 1 pound? Ans. 27 cts.

2. If 8 lbs. of coffee cost $2,04 cts., what is the price of one pound? Ans. 25½ cts.

3. Bought 29 yds. of fine linen for $65,25 cts., what was the price per yard? Ans. $2,25.

4. Paid $58,75 cts. for 235 yds. of muslin, what was it per yard? Ans. 25 cts.

5. A piece of cloth containing 72 yds. cost $450, what was it per yard? Ans. $6,25.

Questions.

How do you perform division of Federal Money? How do you proceed when a remainder occurs?

PROMISCUOUS EXERCISES IN THE PRECEDING RULES.

1. Bought 18 barrels of potatoes, each containing 3 bushels, at 25 cts. a bushel, what did they cost? Ans. $13,50.

2. A farmer sold 30 bushels of rye at 87 cts. a bushel, 30 bushels of corn at 53 cts. a bushel; 8 bushel of beans at $1,25 cts. a bushel; 2 yoke of oxen at $62 a yoke; 10 calves at $4 a piece; 15 barrels of cider at $2,37½ a barrel, what was the amount of the whole? Ans. $251,62½.

3. What will be the price of four bales of goods, each bale containing 60 pieces, and each piece 49 yards, at 37½ cents a yard? Ans. $4410.

4. Add $324,43½ cts. $208,09½ cts. and $507,90½ cts. together, and divide the sum by 2, and what will be the result? Ans. $520,21¾.

5. Divide 400 dollars, equally, among 20 persons. What will be the portion of each person? Ans. $20.

6. Divide 1728 dollars, equally among 12 persons. What does each one of them share? Ans. $144.

7. If 240 bushels cost 420 dollars; what is the cost of one bushel at the same rate? Ans. $1.75.

REDUCTION.

Reduction is the changing of a sum, or quantity, from one denomination to another, without altering the value.

Case 1.

To reduce a sum, or quantity, to a lower denomination than its own.

Rule.—Multiply the sum, or quantity, by that number of the lower denomination which makes one of its own.

If there are one or more denominations between the denomination of the given sum, and that to which it is to be changed, first change it to the next lower than its own; then to the next lower, and so on to the denomination required.

DRY MEASURE.

TABLE.

2 pints (pts.) make		1 quart,	qt.
8 quarts	-	1 peck,	pc.
4 pecks	-	1 bushel,	bu.

Note.—This measure is used for measuring grain, salt, fruit, &c.

EXAMPLES.

Note.—1. To reduce bushels to pecks, multiply by 4, because each bushel has 4 pecks in it.

1. Reduce 23 bushels to pecks.

bu.
23
4
—
Amt. 92 pecks.

2. Reduce 35 bushels to pecks. Amt. 140 pecks.

Note.—2. To reduce pecks to quarts, multiply by 8, because each peck has 8 quarts in it.

3. Reduce 27 pecks to quarts.

pe.
27
8
———
Amt. 216 quarts.

4. Reduce 43 pecks to quarts. Amt. 344 quarts.

NOTE.—3. To reduce quarts to pints multiply by 2, because each quart has 2 pints in it.

5. Reduce 43 quarts to pints.

qt.
43
2
—
Amt. 86 pints.

6. Reduce 32 quarts to pints. Amt. 64 pints.

Reduce 34 bushels to pints.

bu.	
34	
4	Multiply the bushels by 4 to bring them to pecks.
136	
8	Multiply the pecks by 8 to bring them to quarts.
1088	
2	And multiply the quarts by 2 to bring them to pints.

Amt. 2176 pints.

EXERCISES.

7. Reduce 56 pecks to pints. Amt. 896 pints.
8. Reduce 47 bushels to quarts. Amt. 1504 qt.
9. Reduce 85 bushels to pints. Amt. 5440 pt.
10. Reduce 63 pecks to quarts. Amt. 504 qt.
11. Reduce 132 bushels to quarts. Amt. 4224 qt.
12. Reduce 234 bushels to pints. Amt. 14976 pt.

NOTE.—4. When several denominations occur, reduce the highest denomination to the next lower one, and this again to the next lower, and so on; observing to add the amount of each denomination, the number there is of that denomination in the given sum.

EXAMPLES.

1. Reduce 23 bushels, 3 pecks, 5 quarts, 1 pint, to pints.

```
bu.  pe.  qt.  pt.
23 - 3 - 5 - 1
 4
---
92
 3
---
95
 8
---
760
  5
---
765
  2
---
1530
   1
---
1531 amt.
```

Multiply the bushels by 4 to bring them to pecks, and add the 3 pecks to the amount, which makes 95 pecks.

Multiply the pecks by 8 to bring them to quarts, and add the 5 quarts, which makes 765 quarts.

Multiply the quarts by 2 to bring them to pints, and add the 1 pint which makes 1531 pints.

Or thus:

```
bu.  pe.  qt.  pt.
23   3 - 5 - 1
 4
---
95
 8
---
765
  2
---
1531 Amt. as before.
```

Multiply by 4 as above; add the 3, and set down the amount, &c.

EXERCISES.

1. Reduce 13 bushels, 2 pecks, 7 quarts, 1 pint to pints. Amt. 879 pints.

2. Reduce 24 bushels, 3 pecks, 1 quart to quarts. Amt. 793 qt.

3. Reduce 7 bushels, 3 pecks to quarts. Amt. 248 qt.

4. Reduce 3 pecks, 2 quarts to pints. Amt. 52 pt.

5. Reduce 7 quarts, 1 pint, to pints. Amt. 15 pt.

6. Reduce 32 bushels, 0 pecks, 1 quart to pints. Amt. 2050.

7. Reduce 5 bushels, 1 peck, 0 quarts, 1 pint to pints. Amt. 337 pt.

8. Reduce 43 bushels, 1 peck to pints. Amt. 2768 pt.

Questions.

What is reduction?

For what is case first used?

How do you reduce a sum to a lower denomination than its own?

How do you reduce bushels to pecks?

Why do you multiply by 4?

How do you reduce pecks to quarts?

Why do you multiply by 8?

How do you reduce quarts to pints?

How do you reduce bushels to pints?

AVOIRDUPOIS WEIGHT.

TABLE.

16	drams (dr.) make	1 ounce,	oz.
16	ounces -	1 pound,	lb.
28	pounds -	1 quarter of a cwt.	qr.
4	quarters, (or 112 lb.)*	1 hundred weight,	cwt
20	hundred weight	1 ton,	T.

NOTE.—By this weight are weighed, tea, sugar, coffee, flour and other things subject to waste, and all the metals, except silver and gold.

* The gross hundred weight of 112 pounds is nearly out of use; the decimal hundred weight of 100 pounds is taking its place.

EXAMPLES.

1. Reduce 23 tons to hundred weight.

```
      tons.
       23
       20
      ----
Amt. 460 cwt.
```

2. Reduce 34 hundred weight to quarters.

```
      cwt.
       34
        4
      ----
Amt. 136 quarters.
```

3. Reduce 42 quarters to pounds.

```
      qrs.
       42
       28
      ----
      336
      84
      ----
Amt. 1176 pounds.
```

4. Reduce 73 pounds to ounces.

```
      lbs.
       73
       16
      ----
      438
      73
      ----
Amt. 1168 ounces.
```

5. Reduce 54 ounces to drams.

```
      oz.
      54
      16
      ----
      324
      54
      ----
Amt. 864 drams.
```

6. Reduce 35 tons to drams.

```
          tons.
           35
           20
          ----
          700 cwt.
            4
         -----
         2800 qr.
           28
        ------
        22400
        5600
        ------
        78400 lb.
           16
       -------
       470400
       78400
      --------
      1254400 oz.
           16
      --------
      7526400
     1254400
     ---------
Amount. 20070400 drams.
     ---------
```

EXERCISES.

7 Reduce 24 pounds to drams. Amt. 6144 dr.

8. Reduce 36 hundred weight to pounds. Amt. 4032 lb.

9. Reduce 73 quarters to ounces. Amt. 32704 oz.

10. Reduce 2 tons to pounds. Amt. 4480 lb.

11 Reduce 4 tons to drams. Amt. 2293760 dr.

12. Reduce 3 tons, 13 cwt., 2 qu., 14 lbs., to pounds.

```
T. cwt. qr. lb.
3 - 13 - 2 - 14
20
—
60
13
—
73
 4
——
292
  2
——
294
 28
———
2352
588
———
8232
  14
———
8246 pounds.
```

Or thus:

```
T. cwt. qr. lb.
3 - 13 - 2 - 14
20
—
73
 4
——
294
 28
———
2366
588
———
8246 pounds
```

13. Reduce 2 tons. 15 cwt. 2 qr. to quarters. Amt. 222 qr.

14. Reduce 3 tons. 25 lb. to pounds. Amt. 6745 lb.

15. Reduce 5 cwt. 3 qr. 14 lb. to ounces. Amt. 10528 oz.

16. Reduce 2 cwt. 2 qr. 14 ounces to drams. Amt. 71,904 dr.

TROY WEIGHT.

TABLE.

24 grains (gr.) make	1 pennyweight,	dwt.
20 pennyweights -	1 ounce,	oz.
12 ounces -	1 pound,	lb.

NOTE.—By this weight, jewels, gold, silver, and liquors, are weighed.

EXAMPLES.

1. Reduce 32 pounds to ounces.

```
       lb.
       32
       12
       ---
Amt.  384 ounces.
```

2. Reduce 23 ounces to pennyweights.

```
       oz.
       23
       20
       ---
Amt.  460 dwt.
```

3. Reduce 43 pennyweights to grains.

```
      dwt.
       43
       24
      ----
      172
      86
      ----
Amt  1032 grains.
```

4. Reduce 53 pounds to grains.

```
         lbs.
          53
          12
         ---
         636
          20
       -----
       12720
          24
       -----
       50880
      25440
      ------
Amt. 305280 grains.
```

EXERCISES.

1. Reduce 24 ounces to grains. Amt. 11520 gr.
2. Reduce 32 pounds to pennyweights. Amt. 7680 dwt.
3. Reduce 132 pounds to ounces. Amt. 1584 oz.
4. Reduce 234 ounces to grains. Amt. 112320 gr.

5. Reduce 463 pounds to grains. Amt. 2666880 gr.

6. Reduce 47 pounds, 10 ounces, 15 pennyweights to pennyweights. Amt. 11495 dwt.

7. Reduce 5 pounds, 6 ounces, 4 pennyweights, 20 grains to grains. Amt. 31796 gr.

APOTHECARIES WEIGHT.

TABLE.

20 grains (gr.) make	1 scruple,	sc. ℈
3 scruples	-	1 dram,	dr. ʒ
8 drams	-	1 ounce,	oz. ℥
12 ounces	-	1 pound,	lb.

NOTE.—By this weight apothecaries mix their medicines, but they buy and sell by Avoirdupois Weight.

EXERCISES.

1. Reduce 32 pounds to ounces. Amt. 384 oz.
2. Reduce 43 ounces to drams. Amt. 344 dr.
3. Reduce 27 drams to scruples. Amt. 81 sc.
4. Reduce 37 scruples to grains. Amt. 740 gr.
5. Reduce 28 pounds to drams. Amt. 2688 dr.
6. Reduce 36 ounces to scruples. Amt. 864 sc.
7. Reduce 27 drams to grains. Amt. 1620 gr.
8. Reduce 23 pounds to grains. Amt. 132480 gr.
9. Reduce 3 pounds, 5 ounces, 2 scruples to scruples. Amt. 986. sc.
10. Reduce 7 ounces, 5 drams, 14 grains to grains. Amt. 3674 gr.
11. Reduce 27 pounds, 7 ounces, 2 drams, 1 scruple, 2 grains, to grains. Amt. 159022 gr.

CLOTH MEASURE.

TABLE.

4 nails (na.) make		1 quarter of a yard,	qr.
4 quarters	-	1 yard,	yd.
3 quarters	-	1 Ell Flemish,	E. Fl
5 quarters	-	1 Ell English,	E. E.
6 quarters	-	1 Ell French,	E. Fr.

NOTE.—By this measure cloth, tapes, linen, muslin, &c., are measured.

EXERCISES.

1. Reduce 24 yards to quarters. Amt. 96 qr.
2. Reduce 32 quarters to nails. Amt. 128 na.
3. Reduce 27 yards to nails. Amt. 432 na.
4. Reduce 46 Flemish ells to quarters. Amt. 138 qr.
5. Reduce 27 English ells to quarters. Amt. 135 qr.
6. Reduce 34 French ells to quarters. Amt. 204 qr.
7. Reduce 45 Flemish ells to nails. Amt. 540 na.
8. Reduce 36 English ells to nails. Amt. 720 na.
9. Reduce 54 French ells to nails. Amt. 1296 na.
10. Reduce 13 yards, 3 quarters to quarters. Amt. 55 qr.
11. Reduce 3 quarters, 2 nails, to nails. Amt. 14 na.
12. Reduce 24 yards, 2 nails to nails. Amt. 386 na.
13. Reduce 13 E. ells, 2 qrs., 3 nails to nails. Amt. 271 na.

LONG MEASURE.

TABLE.

12 inches (in.) make	1 foot,		ft.
3 feet	-	1 yard,	yd.
5½ yards	-	1 Rod, Pole, or Perch,	p.
40 poles	-	1 Furlong.	
8 Furlong	-	1 Mile.	
3 Miles	-	1 League.	
60 Geographic, or, 69½ Statute Miles		1 degree.	

NOTE.—This *measure is used* for length and distances.

A Hand is a measure of four inches, and is used in measuring the height of horses.

A Fathom is 6 feet, and is chiefly used in measuring the depth of water.

EXERCISES.

1. Reduce 23 leagues to miles. Amt. 69 m.
2. Reduce 43 miles to furlongs. Amt. 344 f.
3. Reduce 27 furlongs to poles. Amt. 1080 p.
4. Reduce 56 poles to yards. Amt. 308 yd.
5. Reduce 132 yards to feet. Amt. 396 ft.
6. Reduce 76 feet to inches. Amt. 912 in
7. Reduce 24 miles to poles. Amt. 7680 p.
8. Reduce 32 furlongs to yards. Amt. 7040 yd.
9. Reduce 86 poles to inches. Amt. 17028 in.
10. Reduce 26 leagues to yards. Amt. 137280 yd.
11. Reduce 52 miles to feet. Amt. 274560 ft.
12. Reduce 5 leagues to inches. Amt. 950400 in.
13. Reduce 24 degrees to statute miles. Amt. 1668 m.
14. Reduce 12 miles, 3 furlongs, 25 poles to poles. Amt. 3985 po.
15. Reduce 14 leagues, 2 furlongs to poles. Amt. 13520 po.
16. Reduce 3 leagues, 2 miles, 6 furlongs, 18 poles to yards. Amt. 20779 yds.

LAND, OR SQUARE MEASURE.

TABLE.

144 square inches make		1 square foot,	ft.
9 square feet	-	1 square yard,	yd.
30¼ square yards	-	1 square perch,	p.
40 square perches	-	1 rood,	r.
4 roods	-	1 acre,	a.

NOTE.—This measure is used to ascertain the quantity of lands, and of other things having length and breadth to be estimated.

EXERCISES.

1. Reduce 27 acres to roods. Amt. 108 r.
2. Reduce 53 roods to perches. Amt. 2120 p.

3. Reduce 28 perches to square yards.
Amt. 847 sq. yds.
4. Reduce 36 square yards to square feet. 324 ft.
5. Reduce 27 square feet to square inches.
Amt. 3888 in.
6. Roduco 34 acres to perches. Amt. 5440 p.
7. Reduce 42 roods to square yards.
Amt. 50820 sq. yds.
8. Reduce 24 square perches to square feet.
6534 ft.
9. Reduce 32 roods to square feet. Amt. 348480 ft.
10. Reduce 23 acres to square inches.
Amt. 144270720 sq. in.
11. Reduce 11 acres, 2 roods, 19 perches to perches.
Amt. 1859 p.
12. Reduce 17 acres, 3 roods to perches.
Amt. 2840 p.
13. Reduce 12 acres, 2 roods, 12 perches to square yards. Amt. 60863 sq. yd.

CUBIC, OR SOLID MEASURE.

TABLE.

1728 cubick inches make	1 cubic foot
27 feet	1 cubic yard
40 feet of round timber, or 50 fect of hewn timber,	1 Ton or load
128 solid feet	1 Cord of wood

NOTE.—This measure is employed in measuring solids, having length, breadth, and thickness to be estimated.

EXERCISES.

1 Reduce 29 cords of wood to cubick feet.
Amt. 3712 c. f.
2 Reduce 32 cubic yds. to feet. Amt. 864 c. f.
3 Reduce 23 cubic feet to inches. Amt. 39744 c. in.
4 Reduce 32 cubic yds. to inches. Amt. 1492992 c. in.
5 Reduce 2 cords of wood to inches. Amt. 442368 c. in.
6 Reduce 3 cords, 10 feet to feet. Amt. 394 ft
7 Reduce 1 cord, 3 feet, 136 inches to inches.
Amt. 226504 in.

LIQUID MEASURE.

TABLE.

4 gills make	1 pint	pt.
2 pints (pts)	1 quart	qt.
4 quarts	1 gallon	gal.
42 gallons	1 tierce	te.
63 gallons	1 hogshead	hhd.
2 hogsheads	1 pipe or butt	pi.
2 pipes	1 tun.	T—

NOTE.—This measure is employed in measuring cider, oil, beer, &c.

EXERCISES.

1 Reduce 23 tuns to pipes. Amt. 46 pi.
2 Reduce 43 pipes to hogsheads. Amt. 86 hhd.
3 Reduce 34 hogsheads to gallons. Amt. 2142 gal.
4 Reduce 27 tierces to gallons. Amt. 1134 gal.
5 Reduce 53 gallons to quarts. Amt. 212 qt.
6 Reduce 724 quarts to pints. Amt. 1448 pt.
7 Reduce 37 pints to gills. Amt. 148 g.
8 Reduce 12 pipes to gallons. Amt. 1512 gal.
9 Reduce 4 hogsheads to quarts Amt. 1008 qt.
10 Reduce 32 gallons to gills. Amt. 1024 g.
11 Reduce 2 tuns to gills. Amt. 16128 gills
12 Reduce 32 gals 3 qts. to pints. Amt. 262 pt.
13 Reduce 2 hogsheads, 27 gals. 3 qts to quarts. Amt. 615 qt.
14 Reduce 3 tons, 1 hogshead, 15 gals. 1 qt to pints. Amt. 6674 pt.

MOTION, OR CIRCLE MEASURE.

TABLE.

60 seconds (″sec) make	1 minute	′min.
60 minutes	1 degree	° deg.
30 degrees	1 sine	. sin.
12 sines (or 360 degrees)	1 revolution	

NOTE.—This measure is employed by astronomers, navigators, &c.

EXERCISES.

1 Reduce 5 sines to degrees. Amt. 150°

2 Reduce 8 degrees to minutes. Amt. 480^{1}

3 Reduce 6 minutes to seconds. Amt. 360 sec.

4 Reduce 12 sines to seconds. Amt. 1296000 sec.

5 Reduce 3 sines 15 degrees to minutes. Amt. 6300 min.

TIME.

TABLE.

60 seconds (sec) make	1 minute	min.
60 minutes	1 hour	H.
24 hours	1 day	
7 days	1 week	
12 months (or 365 days)	1 year.	

Note.—The true year, according to the latest and most accurate observations, consists of 365 d. 5 h. 48 m. and 58 sec: this amounts to nearly 365¼ days. The common year is reckoned 365 days, and every fourth or leap year one day more on account of the fraction omitted each year, which being put together, every fourth year is added to it, making leap year 366 days.

The year is divided into 12 months as follows.

The fourth, eleventh, ninth and sixth,
Have thirty days to each affixed,
And every other thirty-one,
Except the second month alone,
Which has but twenty-eight in fine,
Till leap year gives it twenty-nine.

OR THUS:

Thirty days hath September,
April, June, and November,
February hath twenty-eight alone,
And each of the rest has thirty one.

When the year can be divided by four, without a remainder, it is bissextile, or leap year.

EXERCISES.

1 Reduce 42 years to months. Amt. 504 m.
2 Reduce 23 days to hours. Amt. 552 h.
3 Reduce 36 hours to minutes. Amt. 2160 min.
4 Reduce 25 minutes to seconds. Amt. 1500 sec.
5 Reduce 14 days to minutes. Amt. 20160 min
6 Reduce 52 hours to seconds. Amt. 187200 sec.
7 Reduce 13 weeks to hours. Amt. 2184 h.
8 Reduce 12 weeks to minutes. Amt. 120960 min.
9 Reduce 3 years to minutes, allowing 365 days to each year. Amt. 1576800 min.
10 Reduce 15 years and 6 months to months. Amt. 186 m.
11 Reduce 4 weeks, 3 days, 22 hours, to hours. Amt. 766 h.
12 Reduce 7 years, 24 days, 43 minutes, to seconds. Amt. 222828180 sec.

STERLING MONEY.

TABLE.

4 farthings (qr)	make	1 penny	d.
12 pence		1 shilling	s.
20 shillings		1 pound	

Farthings are usually written as fractions of a penny, thus ·

¼ one farthing
½ two farthings or a half penny.
¾ three farthings.

EXERCISES.

1 Reduce 14 pounds to shillings. Amt. 280 s.
2 Reduce 23 shillings to pence. Amt. 276 d.
3 Reduce 34 pence to farthings. Amt. 136 qr.
4 Reduce 4 pounds to pence. Amt. 960 d.
5 Reduce 13 shillings to farthings. Amt. 624 qr.
6 Reduce 16 pounds to farthings. Amt. 15360 qr.
7 Reduce 13 pounds 14 shillings, to pence. Amt. 3288 d.
8 Reduce 3 pounds 15 shillings 6 pence to farthings. Amt. 3624 qr.

FEDERAL MONEY.

TABLE.

10 mills make	1 cent
10 cents	1 dime
10 dimes	1 dollar
10 dollars	1 eagle

EXERCISES.

1 Reduce 5 eagles to cents. Amt. 5000 ct.
2 Reduce 3 dollars to mills. Amt. 3000 m.
3 Reduce 15 dimes to cents. Amt. 150 ct.
4 Reduce 3 eagles, 5 dollars to cts. Amt.3500 ct.
5 Reduce 7 dollars, 3 dimes, 6 cents, to mills. Amt. 7360 m.

As eagles, dimes and mills are not used in accounts, they will generally be omitted in the subsequent exercises of this work.

4 fourths, or 3 thirds, or 2 halves, make 1 cent.
100 cents - - - 1 dollar.

6 Reduce 125 cents to halves of a cent. Amt. 250 halves.
7 Reduce 32 cents to fourths of a cent. Amt. 128 fourths.
8 Reduce 23 dollars to cents. Amt. 2300 ct.
9 Reduce 25 dollars 15 cents to cents. Amt. 2515 ct.
10 Reduce 15 dollars 37½ cents to halves of a cent. Amt. 3075 halves.
11 Reduce 21 dollars 15 cents to thirds of a cent. Amt. 6345 thirds.
12 Reduce 5 dols. 37½ cents to fourths of a cent. Amt. 2150 fourths.
13 Reduce 15 dollars 33⅓ cts. o thirds of a cent. Amt. 4600 thirds.

NOTE. To reduce dollars to cents annex two cyphers: thus 53 dollars are 5300 cents.

To reduce dollars and cents to cents, place them together without any separating point, and the amount will be cents. Thus 35 dollars 24 cents are 3524 cents.

Questions.

For what purpose is Dry measure used?
For what is Avoirdupois weight used?
For what is Troy weight employed?
For what is Apothecaries weight employed?
For what is Cloth measure employed?
For what is Long measure used?
For what is Land or Square measure used?
For what is Cubick measure employed?
For what is Liquid measure employed?
For what is Sterling currency used?
For what is Federal currency used?

Case 2.

To reduce a sum or quantity to a HIGHER *denomination than its own*

Rule.—Divide the sum or quantity by that number of its own denomination which makes one of the denomination to which it is to be changed.

When there are one or more denominations between the denomination of the given sum and that to which it is to be changed; first change it to the next higher than its own, and then to the next higher, and so on.

Remainders are always of the same denominations as the sums divided.

DRY MEASURE.

EXAMPLES.

1 Reduce 25 pints to quarts. *pts.*

Note.—Divide by 2, because every 2 pints make one quart. In 25 are 12 two's and 1 over, that is 12 quarts and 1 pint.

2)25
——
qt.12—1pt

2 Reduce 43 quarts to pecks. *qt.*

Divide by 8, because every 8 qts. make 1 peck. In 43 are 5 eights and 3 over, that is 5 pecks and 3 quarts.

8)43
——
pe. 5—3qt

3 Reduce 26 pecks to bushels. *bu.*
Divide by four because every 4 pecks make 1 bushel. In 26 are 6 fours and 2 over; that is 6 bushels and 2 pecks.

```
 4)26
 ——
bu. 6—2 pecks
```

4 Reduce 359 pints to bushels. *pt.*
Divide pints by 2, brings them to quarts; divide quarts by 8, brings them to pecks, and divide pecks by 4 brings them to bushels.

```
2)359
 ——
8)179—1 pt.
 ——
4) 22—3 qt.
————————
5 b. 2 p. 3 qt. 1 pt.
```

5 Reduce 81 quarts to bushels. A. 2 bu. 2 pe. 1 qt.
6 Reduce 134 pints to pecks. 8 pe. 3 qt.
7 Reduce 194 pints to bushels. 3 bu. 0 pe. 1 qt.

Questions.

What is reduction?

For what is case second used?

How do you reduce a sum to a higher denomination than its own?

When there are one or more denominations between the denomination of the given sum and the one to which you wish to reduce it, how do you proceed?

Of what denomination is the remainder always?

How do you bring pints to quarts?

How do you bring quarts to pecks?

How do you bring pecks to bushels?

How do you bring pints to bushels?

AVOIRDUPOIS WEIGHT.

1 Reduce 65 cwt. to tons. Result 3 tons 5 cwt.
2 Reduce 27 quarters to cwt. Res. 6 cwt. 3 qr.
3 Reduce 109 pounds to qr. Res. 3 qr. 25 lb.
4 Reduce 123 ounces to pounds. Res. 7 lb. 11 oz.
5 Reduce 234 drams to ounces. Res. 14 oz. 10 dr.
6 Reduce 4274 drams to pounds. Res. 16 lb. 11 oz. 2 dr.
7 Reduce 175 quarters to tons. Res. 2 tons 3 cwt. 3 qr.
8 Reduce 6745 pounds to tons. Res. 3 tons 25 lb.

TROY WEIGHT.

1 Reduce 378 ounces to pounds. Result, 31 lbs. 6 oz.
2 Reduce 235 pennyweights to ounces. Res. 11 oz. 15 dwt.
3 Reduce 748 grains to pennyweights. Res. 31 dwt. 4 grains.
4 Reduce 678 pennyweights to pounds. Res. 2 lbs. 9 oz. 18 dwt.
5 Reduce 732 grains to ounces. Res. 1 oz. 10 dwt. 12 grains.
6 Reduce 14752 grains to pounds. Res. 2 lbs. 6 oz. 14 dwt. 16 gr.

APOTHECARIES WEIGHT.

1 Reduce 432 ounces to pounds. Res. 36 lbs.
2 Reduce 782 drams to ounces. Res. 97 oz. 6 dr.
3 Reduce 91 scruples to drams. Res. 30 dr. 1 scr.
4 Reduce 192 grains to scruples. Res. 9 sc. 12 gr.
5 Reduce 256 scruples to ounces. Res. 10 oz. 5 dr. 1 scr.
6 Reduce 12660 grains to pounds. Res. 2 lb. 2 oz. 3 drs.

CLOTH MEASURE.

1 Reduce 60 quarters to yards. Res. 15 yds.
2 Reduce 60 quarters to English ells. Res. 12 E. ells.
3 Reduce 60 quarters to French ells. Res. 10 Fr. ells.
4 Reduce 60 quarters to Flemish ells. Res. 20 Fl. ells.
5 Reduce 52 nails to quarters. Res. 13 qr.
6 Reduce 123 nails to yards. Res. 7 yds. 2 qr. 3 na.
7 Reduce 543 nails to English ells. Res. 27 ells. 0 qr. 3 nails.

LONG MEASURE.

1 Reduce 36 miles to leagues. Res. 12 l.
2 Reduce 75 furlongs to miles. Res. 9 m. 3f.

3 Reduce 295 poles to furlongs. Res. 7 f. 15 p.
4 Reduce 286 yards to poles. Res. 52 p.
5 Reduce 365 feet to yards. Res. 121 yds. 2 ft.
6 Reduce 759 inches to feet. Res. 63 ft. 3 in.
7 Reduce 253 inches to yards. Res. 7 yds. 0 ft 1 inch.
8 Reduce 2792 poles to leagues. Res. 2 l. 2 m. 5 f. 32 p.

SQUARE MEASURE.

1 Reduce 287 roods to acres. Result 71 a. 3 r.
2 Reduce 245 perches to roods. Res. 6 r. 5 p.
3 Reduce 756 square feet to yards. Res. 84 yds.
4 Reduce 4731 square yards to perches.
Res. 156 p. 12 yds.

```
        Yds.
30¼ |  4731 |
 4  |     4 |
 ---  -----
121 | 18924 | 156 p.
    | 121
      ---
      682
      605
      ---
       774
       726
       ---
    4 | 48 Rem.
      ---
       12 yards.
```

Bring the 30¼ yards and the 4731 yards both to fourths, and divide. The remainder 48, is fourths of a yard; divided by four, brings it to yards, the true remainder.

5 Reduce 3575 square inches to feet.
Res. 24 feet 119 inches.
6 Reduce 1728 square perches to acres.
Res. 10 a. 3 r. 8 p.

CUBIC MEASURE.

1 Reduce 789 cubic feet to cords. Result, 6 c. 21 ft.
2 Reduce 343 cubic feet to yards. Res. 12 yds. 19 ft.
3 Reduce 9386 cubic inches to feet. Res. 5 ft. 746 in.
4 Reduce 703539 cubic inches to cords.
Res. 3 c. 23 ft. 243 in.

LIQUID MEASURE.

1 Reduce 25 pipes to tuns. Res. 12 T. 1 P.
2 Reduce 34 hogsheads to pipes. Res. 17 P.
3 Reduce 1575 gallons to hogsheads. Res. 25 hhds.
4 Reduce 163 quarts to gallons. Res. 40 gal. 3 qt.
5 Reduce 6048 pints to tuns. Res. 3 tuns.

MOTION.

1 Reduce 1440 seconds to minutes. Result, 24 min.
2 Reduce 720 minutes to degrees. Res. 12 deg.
3 Reduce 342 degrees to sines. Res. 11 sines 12 deg.
4 Reduce 443907 seconds to sines.
Res. 4 sines 3 deg. 18 min. 27 sec.

TIME.

1 Reduce 1800 seconds to minutes. Result, 30 m.
2 Reduce 720 minutes to hours. Res. 12h.
3 Reduce 744 months to years. Res. 62 yrs.
4 Reduce 4649 minutes to days. Res. 3d. 5h. 29 min.
5 Reduce 48888 minutes to weeks.
Res. 4w. 5d. 22hrs. 48 min.

STERLING MONEY.

1 Reduce 78 shillings to pounds. Res. 3£. 18*s.*
2 Reduce 93 pence to shillings. Res. 7*s.* 9*d.*
3 Reduce 39 farthings to pence. Res. 9*d.* 3*qr*
4 Reduce 656 pence to pounds. Res. 2£ 14*s.* 8*d.*
5 Reduce 781 farthings to shillings. Res. 16*s.* 3*d.* 1*qr.*
6 Reduce 6529 farthings to pounds.
Res. 6£. 16*s.* 0*d.* 1 *qr.*

FEDERAL MONEY.

1 Reduce 250 halves to cents. Res. 125 cts.
2 Reduce 128 fourths to cents. Res. 32 cts.
3 Reduce 2343 cents to dollars. Res. 23 dol. 43 cts.
4 Reduce 1537½ cents to dollars. Res. 15 dol. 37½ cts.
5 Reduce 6150 half cents to dollars. Res. 30 dol. 75 cts.

NOTE.—To reduce cents to dollars, separate two figures on the right hand for cents; those on the left will be dollars.

PROMISCUOUS EXERCISES.

1 How many bushels in 738 quarts?
Ans. 23 bushels 2 quarts.

2 In 7 bushels, how many pints? Ans. 448 pints.

3 How many cwt. in 5356 ounces?
Ans. 2 cwt. 3qr. 26lb. 12 oz.

4 How many drams in 3 qr. 23 lb. 14 oz.?
Ans. 27616 drams.

5 How many grains in 9 oz. 14 dwt. 3 gr.?
Ans. 4659 gr.

6 How many pounds in 7432 dwt.? Ans.
30lb. 11 oz. 12 dwt.

7 How many scruples in 15 lb. 1 oz. 6 drams?
Ans. 4362 scr.

8 How many ounces in 216 scruples? Ans. 9 oz.

9 How many furlongs in 2346 yards? Ans.
10f. 26p. 3yd.

10 How many poles in 3 leagues? Ans. 2880 poles.

11 How many yards in 84 nails? Ans. 5 yd. 1 qr.

12 How many perches in 4719 square yards
Ans. 156 perches.

13 How many square yards in one acre?
Ans. 4840 sq. yds.

14 How many hogsheads in 9728 gills?
Ans. 4 hhds. 52 gal.

15 How many pints in 2 pipes? Ans. 2016 pints.

16 How many minutes in 3 days 6 hours? Ans. 4680m.

17 How many hours in 2 weeks and 4 days?
Ans. 432 hours.

18 How many shillings in 27 four-pences? Ans. 9 s.

19 How many cords of wood in 9334 cubic feet?
Ans. 73 cords 20 feet.

20 How many cubic feet in 9 cords? Ans. 1152 feet.

21 How many inches round the globe, which is 360 degrees of 69½ miles each? Ans. 1,585,267,200 inches.
Enumerate the answer.

COMPOUND ADDITION.

Compound Addition is the art of collecting several numbers of different denominations into one sum.

Rule.

Place the numbers so that those of the same denomination may stand directly under each other, observing to set the lowest denomination on the right, the next lowest next, &c.

Then add up the several columns beginning with the lowest denomination: divide the sum by as many of the number of that denomination as it takes to make one of the next; and so on.

Proof.—As in Simple Addition.

DRY MEASURE.

EXAMPLES.

bu.	*pe.*	*qt.*	*pt.*
3	2	7	1
7	3	4	1
3	1	3	1
6	3	2	1
2	2	6	1
24	2	0	1

The first column on the right makes five pints. Five pints make two quarts and leaves one pint. Set down the one pint under the column of pints and carry the two quarts to the column of quarts. The column of quarts with the two quarts added makes twenty four quarts. Twenty-four quarts make three pecks and leave no quarts. Set down 0 under the column of quarts and carry the three pecks to the column of pecks.

The column of pecks with the three pecks added makes fourteen pecks. Fourteen pecks make three bushels and leave two pecks. Set down the two pecks under the column of pecks and carry the three to the column of bushels.

The column of bushels with the three bushels added makes twenty-four bushels. Here set down the whole amount.

bu.	*pe.*	*qt.*	*pt.*
23	3	7	1
34	2	6	1
42	3	5	1
51	1	4	1
23	2	3	1
14	1	2	1
11	3	4	1
202	3	2	1

bu.	*pe.*	*qt.*
4	3	7
5	2	6
6	1	5
7	0	4
8	3	3
4	1	2
3	2	4
40	3	7

bu.	*pe.*	*qt.*	*pt.*
	3	7	1
	2	6	1
	0	5	0
	1	4	1
	3	3	1
	0	2	1
	2	1	1
3	2	7	0

APPLICATION.

1 Add 2 bu. 3 pe.; 7 bu. 3 qt.; 4 bu. 1 pe. 1 pt.; 6 bu. 4 qt. 1 pt.; and 3 pe. 1 qt. together.
Amount 21 bu. 0 pe. 1 qt. 0 pt.

2 Add 3 bu. 2 pe. 3 qt. 1 pt.; 7 bu. 7 qt. 1 pt.; 3 pe. 1 pt.; 4 bu. 5 qt.; 4 bu. 3 pe.; 8 bu. 3 pe. 7 qt. 1 pt. together.
Amt. 29 bu. 2 pe. 0 qt. 0 pt.

3 Add 7 bu. 1 pt.; 3 pe. 7 qt. 1 pt.; 6 qt. 1 pt.; 9 bu. 3 pe. 6 qt. 1 pt.; 3 bu. 3 qt.; 4 bu. 1 pe. Amt. 25 bu. 2 pe.

4 In a wagon load of grain contained in seven sacks, viz: in the first 4 bu. 3 pe. 1 qt.; in the second 5 bu. 7 qt. 1 pt.—the third 3 bu. 1 pe. 1 pt.—fourth, 3 bu. 2 pe. 6 qt.—fifth, 5 bu.—sixth, 4 bu. 1 pe. 1 pt.:—and in the seventh 6 bu. 1 pe. 1 pt. How many bushels?

Questions.

What is *Compound Addition?*

How do you place the numbers to be added?

Do you place the greater or smaller denominations in the right hand column?

Where do you begin the addition?

When the first column is added, how do you proceed with the sum?

When you divide the sum by as many of that denomination as make one of the next; which do you set down, the remainder or the quotient?

What do you do with the quotient?

In what particular does compound addition differ from simple addition?

Do you carry one for *every ten* in compound addition?

Since you do not carry one for every ten, how many do you always carry? A. One for as many of any denomination as make one of the next.

Here the pupil will have something with which to compare *simple addition*, in which he carries one for every ten. This comparison will improve and correct his understanding of the elementary rules.

AVOIRDUPOIS WEIGHT.

	T.	*cwt.*	*qr.*	*lb.*
(1)	15	3	2	15
	4	8	3	9
	82	19	1	10
	163	8	3	17
	34	15	2	24
	300	16	1	19

	T.	*cwt.*	*qr.*	*lb.*	*oz.*	*dr.*
(2)	7	11	2	16	4	13
	15	7	3	8	16	7
	138	19	1	12	8	13
	42	8	3	19	12	4
	357	6	2	8	3	3

APOTHECARIES' WEIGHT.

	℔	℥	ʒ	℈
(1)	6	3	1	2
	19	9	5	1
	182	7	3	2
	57	6	1	0
	40	5	0	0
	306	7	3	2

	℔	℥	ʒ	℈	*gr.*
(2)	84	7	6	0	12
	132	5	0	2	
	16	2	2	2	8
	1427	6	7	0	19
	14	0	6	1	9

TROY WEIGHT.

	lb.	*oz.*	*dwt.*
(1)	47	10	12
	38	8	6
	16	11	4
	7	2	16
	13	9	11
	124	6	9

	lb.	*oz.*	*dwt.*	*gr.*
(2)	185	2	19	20
	56	9	15	6
	1472	11	2	17
	385	0	8	5
	10	8	7	12

CLOTH MEASURE.

	yds.	qr.	na.		E.E.	qr.	na.		E.Fl.	qr.	na.
(1)	75	3	2	(2)	72	3	2	(3)	19	2	3
	163	1	3		536	2	1		728	1	2
	245	2	0		847	1	3		142	0	1
	738	3	1		1453	0	2		816	0	0
	1785	2	3		41	2	0		32	1	2
	3009	1	1								

LONG MEASURE.

	L.	M.	fur.	P.		yd.	ft.	in.
(1)	5	2	4	17	(2)	3	2	11
	16	1	3	10		1	1	9
	72	0	5	24		2	0	8
	526	0	3	12		3	1	10
	834	2	6	34		2	0	4
	38	0	3	12		6	2	7
	1493	2	2	29				

SQUARE MEASURE.

	A.	R.	P.		A.	R.	P.
(1)	39	2	37	(2)	487	2	17
	62	1	17		25	3	28
	68	0	38		67	0	32
	129	3	12		45	1	16
	532	1	18		26	0	29
	832	2	2				

CUBIC MEASURE.

	yd.	ft.	in.
(1)	75	22	1412
	9	26	195
	3	19	1091
	28	15	1110
	49	24	218
	18	17	1225
	186	18	67

	cords.	feet.	in.
(2)	37	119	1015
	9	110	159
	48	127	1071
	8	111	956
	21	9	27
	9	28	1091
	135	122	863

3 In four piles of wood; the first containing 32 feet 149 inches; the second 121 feet 1436 inches; the third 97 feet 498 inches; the fourth 115 feet 1356 inches; how much did the whole amount to?

Ans. 2 cords, 110 ft. 1711 in.

4 In six boat-loads of wood: the first containing 22 cords 114 feet, 987 in.; the second 18 cords, 121 feet, 1436 in.; the third 21 cords, 109 feet, 1629 in.; the fourth 15 cords, 82 feet, 1321 in.; the fifth 16 cords, 98 feet, 1111 in.; the sixth 24 cords, 89 feet, 987 in. How much did they contain?

Ans. 120 c. 105 ft. 559 in.

LIQUID MEASURE.

	T.	hhd.	gal
(1)	18	2	54
	62	1	39
	327	0	4
	46	1	19
	285	3	28
	740	1	18

	hhd.	gal.	qt.	pt.
(2)	385	42	3	1
	27	36	2	0
	132	17	0	0
	729	25	0	0
	173	47	2	1

MOTION.

(1)

°	′	″
17	55	48
1	37	51
28	19	45
19	19	37
67	13	1

(2)

SIN.	°	′	″
1	25	49	51
2	4	21	36
4	19	47	18
1	25	25	39
10	15	24	24

3. Add 5*sin.* 10° 46′ 38″; 11° 37′ 18″; 1*sin.* 17° 12′ 18″; 2*sin.* 52″; 1*sin.* 15° 12′ 23″; and 11° 57′ 29″ together. Ans. 11*sin.* 6° 46′ 58″.

4. Add 45′; 1*sin.* 9° 18″; 14° 21′ 34″; 2*sin.* 8° 13′ 54″; 4*sin.* 7° 12′ 19″; and 47′ 32″ together.
Ans. 8*sin.* 10° 20′ 37″.

TIME.

(1)

Y.	M.
71	11
172	9
35	7
4	10
	6
231	7

(2)

we.	d.	h.
3	5	20
2	3	17
3	6	22
0	4	16
0	3	19
11	3	22

(3)

H.	min.	sec.
20	52	40
122	12	35
68	9	17
135	17	12
24	35	28
371	7	12

STERLING MONEY.

(1)

£	s.	d.
2	3	4
7	1	2
9	7	3
5	2	2½
23	13	11½

(2)

£	s.	d.
7	9	4½
13	7	6¾
4	5	2
10	18	10¾

(3)

£	s.	d.
4	6	4
47	19	7
159	5	3
78	6	11¾

4 Add £763 7*s.* 4*d.*; £39 4*s.* 9*d.*; £162 17*s.* 2*d*; £459 15*s.*; £473 12*s.* 8*d* together.
Ans. £1898 16*s.* 11*d.*

5 Add the following sums: viz. £69 18*s*. 7*d*.; £175 2*s*. 6*d*.; £1582 19*s* 4*d*.; £175 13*s*. 9*d*.; £143 13*s*. 8*d*.; and £212 0*s*. 7*d* Ans. £2359 8*s*. 5*d*.

COMPOUND SUBTRACTION.

COMPOUND SUBTRACTION is the art of finding the difference between two numbers consisting of several denominations.

RULE.

Place the numbers, as in compound, addition with the less under the greater: then begin at the right hand denomination and subtract the lower number from the upper, and set down the remainder.

If the upper number of any denomination be less than the lower one, add to the upper one as many as it takes of that to make one of the next; subtract the lower number from the amount and set down the remainder as before.

Proof.—As in simple subtraction.

EXAMPLES.

bu.	*pe.*	*qt.*	*pt*
7	2	4	1
3	1	2	0
4	1	2	1

bu.	*pe.*	*qt.*	*pt.*
42	3	6	1
31	2	3	1
11	1	3	0

bush.	*pe.*	*qt.*
9	2	5
2	3	7
6	2	6

We cannot take 7 quarts from 5 quarts; then borrow 1 from the 2 pecks. One peck has 8 quarts in it: 8 quarts added to the 5 quarts, make 13 quarts. Take 7 qts. from 13 qts. and 6 qts. remain. Set down the 6 qt.

Because I borrowed 1 from the 2 quarts, I must add one to the 3 below it, which makes the lower figure 4. Now 4 pecks from 2 pecks we cannot take: then borrow one bushel from the 9; that bushel has 4 pecks in it; 4 p. and 2 p. make 6 p. Now 4 p. from 6 p. and 2 pecks remain, which set down.

Because I borrowed 1 from the 9, I must add 1 to the figure below it. 1 to 2 make 3. Take 3 from 9 and 6 remain. Set down the 6, and the work is done.

bu.	*pe.*	*qt.*	*pt.*
8	2	7	1
4	3	6	1
3	3	1	0

bu.	*pe.*	*qt.*	*pt.*
8	1	3	0
4	2	5	1
3	2	5	1

bu.	*pe.*	*qt.*
95	3	2
22	0	1

bu.	*pe.*	*qt.*	*pt.*
28	2	2	0
14	3	5	1

APPLICATION.

1. From a granary containing 94 bushels, 2 pecks, 7 quarts, have been taken 43 bush. 3 pe. 5 qr. How much remains? Ans. 50 bush. 3 pe. 2 qt.

2. From a wagon load of corn containing 63 bushels, 3 pecks, 4 qts., have been sold 27 bush. 3 pe. 7 qt. 1 pt. How much remains unsold? Ans. 35 bu. 3 p. 4 qt. 1 pt.

Questions.

What is *compound subtraction?*

In what particular does it differ from simple subtraction?

How do you place the numbers in compound subtraction?

Where do you begin the operation?

When the upper number of any denomination is less than the lower one, how do you proceed?

Do you borrow one from the next?

Do you call the number you borrow, one *ten*, as in simple subtraction?

What do you call it?

Ans. I call it one *peck*, or one *yard*, or one *mile*, as the case may be?

What do you do with it then?

Ans. I reduce it to *quarts*, or to *feet*, or to *furlongs*, &c. according to the nature of the case; then add these to the upper figure on the right, subtract the lower figure from the sum, and set down the remainder.

When you borrow one from the upper figure, why do you add one to the figure below it?

NOTE.—Upon a clear conception of the principles involved in these questions, depends the pupil's correct knowledge of the *science* of Arithmetic.

AVOIRDUPOIS WEIGHT.

	tons	*cwt.*	*qr.*	*tons*	*cwt.*	*qr.*	*lb.*	*cwt.*	*qr.*	*lb.*	*oz.*	*dr.*
From	45	11	3	52	12	3	15	17	0	0	0	0
Take	15	10	2	24	10	0	26	6	3	21	15	9
Rem.	30	1	1	28	2	2	17	10	0	6	0	7

1. Subtract 76 tons, 18 hundred weight, 3 quarters, from 195 tons, 2 hundred weight, 2 quarters.
Ans. 118 tons, 3 cwt. 3 qr.

2. Subtract 14 pounds, 6 ounces, 3 drams from 20 pounds, 2 ounces. Ans. 5 lbs. 11 oz. 13 dr.

APOTHECARIES WEIGHT.

℔	℥	ʒ		℔	℥	ʒ	℈	*gr.*
1090	1	6		48	9	6	1	4
106	2	7		1	10	0	2	8
983	10	7						

3. From 59℔ 1℥ 2ʒ take 53℔ 7℥ 5ʒ. Ans. 5℔ 5℥ 5ʒ.
4. Subtract 14℔ 9℥ 1ʒ from 69℔. Ans. 54℔ 2℥ 7ʒ.

TROY WEIGHT.

lb.	*oz.*	*dwt.*	*gr.*	*lb.*	*oz.*	*dwt.*	*gr.*	*lb.*	*oz.*	*dwt.*	*gr.*
10	6	18	0	8	3	0	2	106	0	0	15
4	0	2	20	2	1	18	6	10	6	2	20
6	6	15	4								

4. Subtract 14*lb.* 6*oz.* 11*dwt.* from 22*lb.* 12*dwt.* 6 *gr.*
Ans. 7*lb.* 6*oz.* 1*dwt.* 6*gr.*

5. From 16*lb.* take 12*lb.* 11*oz.* 10*dwt.* 11*gr.*
Ans. 3*lb.* 0*oz.* 9*dwt.* 13*gr.*

CLOTH MEASURE.

	yds.	*qrs.*	*na.*	*E.E.*	*qrs.*	*na.*	*E. Fl.*	*qrs.*	*na.*
From	71	3	1	42	0	2	51	2	2
Take	14	2	3	19	2	3	42	2	1
Rem.	57	0	2	22	2	3	9	0	1

4. Subtract 95 yards, 3 quarters, 2 nails, from 156 yards, 2 quarters, 3 nails. Ans. 60 yds. 3 qr. 1 nail.

5. Subtract 14 English ells, 1 quarter, 2 nails, from 52 English ells, 3 quarters, 2 nails. Ans. 38 yds. 2 qr.

LONG MEASURE.

	L.	*M.*	*fur.*	*L.*	*M.*	*fur.*	*P.*	*yds.*	*ft.*	*in.*
From	24	1	7	56	1	0	19	6	2	10
Take	18	2	4	10	0	7	20	3	2	7
Rem.	5	2	3	46	0	0	39	3	0	3

4. Subtract 45 miles, 5 furlongs, 20 poles, from 320 miles, 3 furlongs, 36 poles. Ans. 274 M. 6 F. 16 P.

5. Subtract 15 yards, 2 feet, 6 inches, from 36 yards, 1 foot, 11 inches. Ans. 20 yds. 2 ft. 5 inches.

LAND, OR SQUARE MEASURE.

	A.	*R.*	*P.*	*A.*	*R.*	*P.*	*Yds.*	*ft.*	*in.*
From	96	3	36	195	2	2	25	2	72
Take	25	2	39	36	3	1	14	7	10
Rem.	71	0	37	158	3	1	10	4	62

4. Subtract 36 acres, 2 roods, from 900 acres, 3 roods, 16 perches. 864 A. 1 R. 16 P.

5. Subtract 72 acres, from 360 acres, 2 roods, 29 perches. 288 A. 2 R. 29 P.

CUBICK MEASURE.

yds.	*ft.*	*in.*	*cords.*	*ft.*	*in.*
79	11	917	349	97	1250
17	25	1095	192	127	1349
61	12	1550	156	97	1629

1. From a pile of wood containing 432 cords, 27 feet, and 1432 inches, have been hauled 156 cords, 92 feet, 946 inches: how much remains?

Ans. 275 cords, 63 feet, 486 in.

2. From a bank of earth containing 2984 yards, 18 feet, have been taken 1436 yards, 21 feet: what remains? Ans. 1547 yds. 24 feet.

LIQUID MEASURE.

T.	*hhd.*	*gal.*	*qt.*	*pt.*
2	3	50	1	0
1	2	16	3	1
1	1	33	1	1

T.	*hhd.*	*gal.*	*qt.*	*pt.*
100	1	19	2	1
99	1	28	3	1

3. If I purchase 2*hhd.* of wine, and to oblige a friend send him 29*gal.*, what quantity have I left?

Ans. 1*hhd.* 34*gal.*

4. Bought 1 pipe of wine, 4*hhd.* of brandy, 2 barrels of beer; I have since sold 93 gallons of wine, 29 of brandy, 1 barrel of beer: how much of each have I remaining?

Ans. 33*gal.* of wine, 223*gal.* of brandy, and 1 barrel of beer.

MOTION.

°	′	″
79	21	31
41	41	52
37	39	39

Sin.	°	′	″
6	10	12	48
3	8	39	29
3	1	33	19

A circle being 12 sines, how far has the hand of a watch to pass, after having gone through 4 sines, 23° 5′ 29″?

Sin.	°	′	″
12	0	0	0
4	23	15	29

2 A person residing in latitude 27° 32′ 45″ north, wishes to visit a place 52° 24′ 18″ north. How many degrees, minutes, and seconds northward must he travel? Ans. 24° 51′ 33″.

TIME

Y.	M.		w.	d.	ho.	min.	sec.		H.	min.	sec.		Y.	M.
6	9		3	1	3	40	20		16	29	33		18	11
1	6		2	6	2	57	36		7	36	44		9	10
5	3			2	0	42	44							

4. From 900*Y.* take 111*Y.* 6*m.*

Ans. 788*Y.* 6*m.*

5. If I take 1*Y.* 1*M.* from 6*Y.* what space of time will still remain?

Ans. 4*Y.* 11 *M.*

NOTE.—To ascertain the amount of time passed between two events, set down the *year*, *month*, and *day* of the latter event, and place those of the former below it, and subtract.

6. A bond was given 24th July, (7th month) 1809, and paid off 13th August, 1821.

yrs.	*mo.*	*ds.*
1821	8	13
1809	7	24
12	0	20

7. The declaration of independence of the United States passed Congress, 4th July, (7th month) 1776; and the declaration of the late war with Great Britain, 18th June, (6th mo.) 1812. How many years, &c. between them? Ans. 35yr. 11mo. 14d.

STERLING MONEY.

£.	s.	d.		£	s.	d.		£	s.	d.
146	19	10½		47	6	7¾		419	7	6
7	19	9¾		28	5	10½		227	8	9½
139	0	0¾								

4. Subtract £200 9*s.* from £1000 11*s.* 11¾*d.*
Ans. £800 2*s.* 11¾*d.*

5. I have a purse of money containing £1000 2*s.* 4½*d.*: if I take out £60 7*s.* 8¾*d.* what sum will be left?
Ans. £939 14*s.* 7¾*d.*

COMPOUND MULTIPLICATION.

Compound multiplication is the art of multiplying numbers composed of several denominations.

Case 1.

When the multiplier does not exceed 12.

RULE.

Place the number to be multiplied as directed in compound addition; and set the multiplier under the lowest denomination.

Multiply as in simple multiplication, and divide the product of each denomination by as many as it takes of that to make one of the next greater; set down the remainder (if any) and carry the quotient to the product of the next denomination.

Proof.—Double the multiplicand and multiply by half the multiplier.

EXAMPLES.

Bu.	*pe.*	*qt.*	*pt.*
7	2	5	1
			7
53	2	6	1

7 times 1 pint make 7 pints; 2 pt. make 1 qt.; then 7 pt. make 3 qt. and leave 1 pt. Set down the 1 pt. and carry the 3 qt. to the product of the next figure.

7 times 5 qt. make 35 qt. to which add the 3 qt. which make 38 qt.; 8 qt. make one peck; then 38 qt. make 4 pecks and leave 6 qts. Set down the six quarts and carry the 4 pecks.

7 times 2 pecks make 14 pecks; add the 4 pecks, makes 18 pe.; 4 pecks make one bushel; then 18 pe. make 4 bushels and leave 2 pecks. Set down the 2 pecks and carry the 4 bushels.

7 times 7 bushels make 49 bushels; add the 4 bushels, makes 53 bushels, which set down, and the work is done.

Bu.	*pe.*	*qt.*	*pt.*		*Bu.*	*pe.*	*qt.*	*pt.*
9	3	6	1		23	2	5	1
			5					8
49	3	0	1		189	1	4	0

1. In one vessel are contained 29 bushels 2 pecks and 5 quarts: how many in 9 such vessels?

Ans. 266 bu. 3 pe. 5 qt.

2. If one tub will contain 8 bu. 3 pe. 5 qt. how much will 11 such tubs contain? Ans 97 bu. 3 pe. 7 qt.

CASE 2.

When the multiplier exceeds 12, and is the exact product of two factors in the multiplication table.

RULE.

Multiply the given sum by one of the factors, and the product by the other factor.

Proof.—Change the factors.

EXAMPLES.

1. Multiply 3 bushels, 2 pecks, 7 qt. by 24. Product.

bu.	*pe.*	*qt.*		*bu.*	*pe.*	*qt.*
3	2	7		3	2	7
		6				4
22	1	2		14	3	4
		4				6
89	1	0	Proof	89	1	0

OR THUS:

bu.	*pe.*	*qt.*		*bu.*	*pe.*	*qt.*
3	2	7		3	2	7
		3				8
11	0	5		29	3	0
		8				3
89	1	0		89	1	0

2. Multiply 7 bushels, 3 pecks, 5 quarts, by 36.

Product, 284 bu. 2 pe. 4 qt.

3. Multiply 19 bushels, 2 pecks, 3 quarts, bv 42

CASE 3.

When the multiplier exceeds 12, and is NOT *the product of any two factors in the multiplication table.*

RULE.

Multiply by the two factors whose product is the least short of the given multiplier; then multiply the given sum by the number which supplies the deficiency; and add its product to the sum produced by the two factors.

EXAMPLES.

1. Multiply 21 bushels, 1 peck, 7 quarts, by 23. Prod.

bu.	*pe.*	*qt.*		*bu.*	*pe.*	*qt.*	
21	1	7×3		21	1	7×2	
		5	OR THUS:			3	
107	1	3		64	1	5	
		4				7	
429	1	4	product of 20	450	3	3	product of 21
64	1	5	product of 3	42	3	6	product of 2
493	3	1	product of 23	493	3	1	product of 23

2. Multiply 19 bushels, 3 pecks, 7 quarts, by 34.
Product, 678 bu. 3 pe. 6 qt.
3. Multiply 7 bushels, 3 pecks, 4 quarts, by 59.
4. Multiply 9 bushels, 3 pecks, 2 quarts, by 47.
5. Multiply 15 bushels, 1 peck, 7 quarts, by 78.
6. Multiply 12 bushels, 2 pecks, by 92.
7. Multiply 17 bushels, 3 quarts, by 98.
8. How many bushels in 104 sacks, each containing 7 bushels, 2 pecks, 3 quarts?
9. How many bushels of wheat on 125 acres, containing 21 bushels, 3 pecks each?

Case 4.

When the multiplier is greater than the product of any two numbers in the multiplication table.

RULE.

Multiply the given number by 10, as many times less one as there are figures in the multiplier.

Multiply that product by the left hand figure of the multiplier.

Multiply the given sum by the units figure of the multiplier; the product of the first 10 by the *tens* figure of the multiplier; the hundreds product by the *hundreds* figure of the multiplier, and so on, till you have multiplied by all the figures of the multiplier except the left hand one.

Add all the products together, and you have the product required.

EXAMPLES.

1. Multiply 3 bushels, 3 pecks, 1 quart, by 456.

Product 1724 bu. 1 pe.

bu.	pe.	qt.
3	3	1×6
		10
37	3	2×5
		10
378	0	4
		4
1512	2	0
22	2	6
189	0	2
1724	1	0 Product

Because there are 3 figures, multiply 2 times by 10. Multiply that product by the left hand figure (4) of the multiplier. Multiply the given number by the units figure (6) and set the product beneath. Multiply the 10's product by the *tens* figure (5) of the multiplier.

Add the several products.

2. Multiply 53 bushels, 2 pecks, 7 quarts, by 2345.

bu.	*pe.*	*qt.*	
53	2	7×5	
		10	
537	0	6+4	
		10	
5371	3	4+3	
		10	
53718	3	0	
		2	
107437	2	0	product of the 2000
16115	2	4	" " 300
2148	3	0	" " 40
268	2	3	" " 5
125970	1	7	Product of the 2345

NOTE.—Let the pupil try experiments, by multiplying simple numbers in this way.

3. Multiply 72 bushels, 1 peck, 2 quarts, by 4723.
Product, 341531 bu. 3 pe. 6 qt.

4. Multiply 13 bushels, 2 pecks, 4 quarts, by 5124.

Questions.

What is *compound multiplication?*

In what does it differ from simple multiplication?

When the multiplier does not exceed 12, how do you proceed?

How many do you always carry?

How do you prove compound multiplication?

How do you proceed when the multiplier exceeds 12, and is the exact product of two numbers in the multiplication table?

When the multiplier exceeds 12, and is *not* the exact product of any two numbers in the table, how do you proceed?

How do you proceed when the multiplier is greater than the product of any two numbers in the table?

AVOIRDUPOIS WEIGHT.

tons.	*cwt.*	*qrs.*
23	12	3
		4
94	11	0

cwt.	*qr.*	*lb.*	*oz.*	*dr.*
7	3	14	9	6
				6
47	1	3	8	4

tons	*cwt.*	*qr.*
7	15	3
		8

cwt.	*qr.*	*lb.*	*oz.*	*dr.*
7	3	24	12	14
				9

5. Multiply 7 tons, 16 cwt. 3 qr. by 24.
Product, 188 T. 2 cwt.

6. Multiply 3 cwt. 2 qr. 21 lb. 14 oz. by 30.
Product, 110 cwt. 3 qr. 12 lb. 4 oz.

7. Multiply 3 tons, 7 cwt. 2 qr. by 34.
Product, 114 tons 15 cwt.

APOTHECARIES WEIGHT.

℔	℥	ʒ	℈
4	8	2	1
			5
23	5	3	2

℔	℥	ʒ	℈	*gr.*
53	10	0	2	12
				9

℔	℥	ʒ	℈	*gr.*
17	5	6	1	4
				12

TROY WEIGHT.

lb.	*oz.*	*dwt.*
67	5	16
		2
134	11	12

lb.	*oz.*	*dwt.*	*gr.*
43	0	8	10
			4

lb.	*oz.*	*dwt.*	*gr.*
113	6	0	6
			6

4. Multiply 41 lb. 6 oz. 18 dwt. 2 gr. by 7.
Ans. 291 lb. 0 oz. 6 dwt. 14 gr.

5. Multiply 91 lb. 4 oz. 14 dwt. 16 gr. by 8.
Ans. 731 lb. 1 oz. 17 dwt. 8 gr

CLOTH MEASURE.

yd.	*qr.*	*na.*	*E.E.*	*qr.*	*na.*	*E.Fl.*	*qr.*	*na.*	*E.Fr.*	*qr.*	*na.*
20	2	3	37	4	2	18	0	3	14	1	3
		6			8			12			9
124	0	2									

5. If 19 yd. 1 qr. 2 na. be multiplied by 5, what number of yards will there be? Ans. 96 yds. 3 qr. 2 na.

6. Multiply 56 Ells Eng. 3 qr. by 9.
Ans. 509 Ells E. 2 qr.

LONG MEASURE.

deg.	*m.*	*fur.*	*p.*	*l.*	*m.*	*fur.*	*p.*	*m.*	*fur.*	*p.*	*yd.*	*ft.*	*in.*
8	1	3	36	4	2	2	29	18	3	20	1	2	10
			12				7						5
96	17	6	32										

4. Multiply 6 deg. 40 m. 7 fur. by 10.
Ans. 65 deg. 61 m. 2 fur.

5. Multiply 44 m. 6 fur. 20 p. by 7.
Ans. 313 m. 5 fur. 20 p.

LAND, OR SQUARE MEASURE.

a.	*r.*	*p.*	*a.*	*r.*	*p.*	*a.*	*r.*	*p.*
49	2	17	19	3	20	10	0	33
		2			6			9
99	0	34						

4. How many acres will 10 men reap in one day, allowing them 1 acre 3 roods 11 perches each?
Ans. 18 A. 0 R. 30 P.

5. Multiply 63 acres 3 roods 18 perches, by 11.
Ans. 702 A. 1 R. 38 P.

6. How many acres in 15 lots, containing 17 acres, 2 roods, and 20 perches each? Ans. 264A. 1R. 20P.

CUBIC MEASURE.

cords.	*ft.*	*in.*
7	28	1327
		6
43	44	1050

yd.	*ft.*	*in.*
19	23	1421
		8
159	1	1000

cords.	*ft.*	*in.*
21	56	1432
		7

yd.	*ft.*	*in.*
27	13	1291
		9

5 In a pile of wood are 14 cords 92 feet; how much in 24 such piles? Ans. 353 c. 32 ft.

6 In a cellar, are contained 42 yards 25 feet; what are the contents of 23 such cellars? Ans. 987 yd. 8 ft.

LIQUID MEASURE.

hhd.	*gal.*	*qt.*
8	43	2
		4
34	48	0

T.	*hhd.*	*gal.*	*qt.*	*pt.*
1	2	16	3	1
				10

pi.	*hhd.*	*gal.*	*qt.*	*pt.*
4	1	19	3	1
				5

4 Multiply 3 T. 2 hhd. 50 gal. 2 qt. by 8.
Ans. 29 T. 2 hhd. 26 gal. 0 qt.

5 Multiply 4 hhd. 41 gal. 1 pt. by 10.
Ans. 46 hhd. 33 gal. 1 qt. 0 pt.

MOTION.

sin.	°	′
3	27	48
		7
27	14	36

sin.	°	′	″
1	24	48	25
			9
16	13	15	45

3 If a planet move through 2*sin.* 15° 23′ of its orbit in one day; how far will it advance in 8 days.
Ans. 20*sin.* 3° 4′

TIME.

years.	mo.		weeks	d.	h.		d.	h.	min.	sec.
7	8		3	5	23		3	14	25	36
	7				8					9
53	8		30	5	16		32	9	50	24

4 If a man can perform a piece of work in 2 yr. 3 mo., how long would it take him to perform 5 such?

Ans. 11 yr. 3 mo.

5 If a laborer dig a drain in 2 weeks, 3 days, how long a time would he require to dig 9 such drains?

Ans. 21 weeks 6 days.

STERLING MONEY.

£	s.	d.		£	s.	d.		£	s.	d.
246	13	3¾		14	6	0¼		111	11	10½
		11				9				10
2713	6	5¼								

		£	s.	d.				£.	s	d.
4	Multiply	37	6	9½	by 5		Prod.	186	13	11½
5	—	56	8	7¾	by 9		—	507	17	9¾

COMPOUND DIVISION.

Compound Division is the art of dividing a sum which consists of several denominations.

Case I.

When the divisor does not exceed 12.

Rule.

Divide the several denominations of the given sum, one after another, beginning with the highest, and set their respective quotients underneath.

When a remainder occurs, reduce it to the next lower denomination, and add it to the number of the next denomination, and divide the sum as before.

EXAMPLES.

	bu.	*pe.*	*qt.*	*pt.*
7)	25	2	6	1
	3	2	5	1

Here 7 into 25 bu. 3 times and 4 remain. Set down the 3.

Reduce the 4 bushels to pecks, which makes 16 pecks: add 16 pecks to 2 pecks, which make 18 pecks.

Now 7 into 18 pe. 2 times, and leave 4. Set down the 2.

Reduce the 4 pecks to quarts, which makes 32 qts. Add 32 qt. to 6 qt.—makes 38 qt.

7 into 38 qt. 5 times, and 3 remain. Set down the 5.

Reduce the 3 qt. to pt.—makes 6 pt., add 6 pt. to 1 pt. makes 7 pints.

7 into 7, 1 time. Set down the 1, and the work is completed.

	bu.	*pe.*	*qt.*
2)	8	2	6
	4	1	3

	bu.	*pe.*	*qt.*
3)	9	3	6

4 Divide 34 bu. 3 pe. 6 qt. between 9 persons.
Ans. 3 bu. 3 pe. 4 qt.

5 92 bu. 3 pe. belong equally to 7 persons; what is the share of each?
Ans. 13 bu. 1 pe.

Case 2.

When the divisor exceeds 12, and is the exact product of two numbers in the multiplication table.

Rule.

Divide the sum by one of the factors, and the quotient by the other.

Multiply the last remainder by the first divisor, and add the first remainder for the true remainder, as in simple division, note 2.

EXAMPLES.

1 Divide 89 bu. 3 pe. 7 qt. by 28.
Quotient 3 bu. 6 qt. 1 pt.—18 Rem.

```
          bu.  pe.  qt.
    (4 |  89   3   7
28 {   |  ---------
    (7 |  22   1   7—3 Rem.
         ------------
          3    0   6—5 Rem.
                     4 First divisor
                     —
                    20
                     3 First rem.
                     —
      True rem.     23 Quarts
                     2
                     —
                 28)46 pints
                     —
                Pt.  1 and a rem. of 18 pints undivided.
```

3. Divide 78 bushels, 3 pecks, 4 quarts, among 32 persons; what will be the share of each?

Ans. 2 bu. 1 pe. 6 qt. 1 pt. and a remainder of 24 pints undivided.

Case 3.

When the divisor is more than 12, *and is* NOT *the exact product of any two numbers in the multiplication table.*

RULE.

Divide the highest denomination of the given sum, as in case 2, simple division; and reduce the remainder, if any, to the next lower denomination; add the number of that denomination to the result, and divide as before.

EXAMPLES.

1. Divide 77 bushels, 1 peck, 7 quarts, by 23.

Quotient, 3 bu. 1 pe. 3 qt. 1 pt. 13 rem.

EXAMPLES.

```
     bu. pe. qt.bu.pe.qt. pt.
23)79  1  7(3  1  6  1+3 pint remaining.
   69
   --
   10
    4
   --
23)41(1 peck
   23
   --
   18
    8
  ---
23)151(6 quarts
   138
   ---
    13
     2
    --
23)26(1 pint
   23
   --
Rem. 3 pints
```

2. A boat load of corn, containing 4927 bushels, 3 pecks, is owned equally by 29 persons: what is the share of each?

Ans. 169 bu. 3 pe. 5 qt. 1 pt., and a rem. of 1 pint.

Questions.

What is *compound division?*

When the divisor does not exceed 12, how do you proceed?

When a remainder occurs, what do you do with it?

Where the divisor exceeds 12, but is the product of two numbers in the table, how do you perform the operation?

How do you find the *true remainder* in the latter case?

When the divisor is more than 12, and is *not* the product of any two numbers in the table, how do you perform the operation?

AVOIRDUPOIS WEIGHT.

	tons	*cwt.*	*qr.*	*lb.*
6)	37	17	3	27
	6	6	1	9–1 rem.

	lb.	*oz.*	*dr.*
7)	46	12	14
	6	10	15–5 R.

	tons	*cwt.*	*qr.*
8)	92	3	3

	cwt.	*qr.*	*lb.*	*oz.*	*dr.*
9)	75	3	23	14	12

5. A quantity of iron weighing 473 tons, 19 cwt., 3 quarters, is owned equally by 22 persons; what is the share of each?

Ans. 21 T. 10 cwt. 3 qr. 16 lb. 8 oz. 11 dr. Rem. 14 dr.

APOTHECARIES WEIGHT.

	℔	℥	ʒ	℈
4)	23	7	5	1
	5	10	7	1

	℔	℥	ʒ	℈	*gr.*
5)	41	6	7	2	14
	8	3	6	1	2–4 rem.

	℔	℥	ʒ	℈
6)	46	9	1	2

	℔	℥	ʒ	℈	*gr.*
7)	93	7	5	2	14

5. Divide 127℔ 3℥ 6ʒ into 17 equal parcels: how much in each parcel? Ans. 7℔ 5℥ 6ʒ 2℈ 16 gr. 8 rem.

TROY WEIGHT.

	lb	*oz.*	*dwt.*
8)	34	10	15

	lb.	*oz.*	*dwt.*	*gr.*
7)	45	11	16	22

	lb.	*oz.*	*dwt.*
9)	78	9	16

	lb.	*oz.*	*dwt.*	*gr.*
8)	82	7	14	21

CLOTH MEASURE.

yd.	qr.	na.		Ells E.	qr.	na.
5)27	3	1		6)37	3	2
5	2	1		6	1	1+4 rem.

yd.	qr.	na.		Ells E.	qr.	na.
7)45	3	2		8)37	3	1

LONG MEASURE.

l.	m.	f.		m.	f.	p.	yd.
6)37	2	2		7)46	7	17	3
6	0	7		6	5	25	2

yd.	ft.	in.		m.	f.	p.	yd.	ft.
6)53	2	9		7)87	6	23	4	2

5. A traveller has a journey of 946 miles, 6 furlongs, to perform in 26 days; how far must he travel each day? Ans. 36 m. 3 f. 12 p. 8 rem.

LAND, OR SQUARE MEASURE.

A.	R.	P.		A.	R.	P.	yds.
7)37	3	27		9)423	3	28	2
5	1	26+5 Rem.		47	0	16	13+5 Rem.

3. A farm containing 746 acres, 3 roods, 29 poles, is to be divided equally between 9 heirs; what is the share of each? Ans. 82 A. 3 R. 38 P. and 7 rem.

CUBIC MEASURE.

cords	ft.		yd.	ft.	in.
8)97	48		9)148	16	493
12	22		16	13	1398+7R.

3. A boat load of wood, containing 92 cords 87 feet, is to be divided between 3 persons; what is the share of each? Ans. 30 c. 114 ft. 1 rem.

4. A quantity of earth, containing 6987 yards, 25 feet, is to be removed by 29 carters; how much must each remove? Ans. 240 yd. 26 ft.

LIQUID MEASURE.

	tuns.	*hhd.*	*gal.*	
5)	37	3	45	
	7	2	21+3 Rem.	

	hhd.	*gal.*	*qt.*	*pt.*
6)	57	36	3	1
	9	37	2	1+1 Rem.

	tuns	*hhd.*	*gal.*
7)	84	2	32

	hhd.	*gal.*	*qt.*	*pt.*
8)	93	43	3	1

5. A quantity of liquor owned equally by 27 persons, the whole quantity being 431 hhd. 47 gals; what is the share of each? Ans. 15 hhd. 62 gal. 1 qt.; 17 rem.

MOTION.

	Sin.	°	′	″
8)	9	16	45	36
	1	5	50	42

	Sin.	°	′	″
9)	11	23	48	54

TIME.

	yr.	*mo.*
11)	848	10
	77	2

	we.	*da.*	*ho.*	*min.*	*sec*
12)	24	6	20	32	24
	2	0	13	42	42

	yr.	*mo.*
4)	375	8

	da.	*ho.*	*min.*	*sec.*
7)	37	16	28	32

STERLING MONEY.

£	s.	d.		£	s.	d.
6)82	14	6		8)143	7	10
13	15	9		17	18	5¼

£	s.	d.		£	s.	d.
7)78	10	11		9)98	17	1

£	s.	d.	£	s.	d.
19)36	16	3(1	18	9	

6 Divide 113£ 13*s.* 4*d.* by 31. What is the quotient?
Ans. 3£ 13*s.* 4*d.*

7 Divide 189£ 14*s.* by 95. Quotient, 1£ 19s. 11*d.*+

PROMISCUOUS EXERCISES

1 In 35 dollars how many cents? Ans. 3500.

2 How many miles are there in 98 furlongs? Ans. 12M. 2fur.

3 How many weeks are there in 365 days? Ans. 52we. 1 da.

4 In 84 half cents how many cents? Ans. 42 cts.

5 In 8 tons 15 cwt., how many hundred weight? Ans. 175 cwt.

6 How many perches are there in 63 roods? Ans. 2520 square per.

7 How many pounds in 157*s.*? Ans. £7 17*s.*

8 In 175 pecks how many bushels? Ans. 43bu. 3pe.

9 In 7642 cents how many dollars? Ans. $76 42cts

10 In 103 pints how many quarts? Ans. 51qt. 1pt.

11 How many minutes are there in 720 seconds? Ans. 12min.

12 In 7 hogsheads, 33 gallons, how many gallons? Ans. 474 gal.

PROPORTION;

OR,

THE SINGLE RULE OF THREE.

PROPORTION is an *Equality* of RATIOS;*

That is, four numbers are proportional, when the *first* has the same ratio to the *second* as the THIRD has to the FOURTH. Thus, as 12 : 4 : : 24 : 8; or as 4 : 12 : : 8 : 24.

The ratio of 12 to 4 is 3 and the ratio of 24 to 8 is 3. Or, the ratio of 4 to 12 is $\frac{1}{3}$, and the ratio of 8 to 24 is $\frac{1}{3}$. Then

Four numbers are proportional, when the first is as many times the second or the same part of the second, as the third is of the fourth. Or, when the ratio of the first to the second equals the ratio of the third to the fourth.

The two quantities compared are called the TERMS of the ratio: the first is called the ANTECEDENT, and the second the CONSEQUENT. In any series of four proportionals, the first and fourth terms are called the EXTREMES, and the second and third the MEANS. The product of the Means, equals the product of the Extremes. Thus in either series above, $12\times8=96$, and $24\times4=96$.

Now suppose we have the three first terms of a series in proportion, and we wish to find the fourth. Divide the product of the second and third terms by the first, and the quotient will be the fourth term. In this manner let the fourth term be found in each of the following series.

2 : 4 : : 8 is to what?	*Ans.* 16.
3 : 11 : : 9 is to what?	*Ans.* 33.
4 : 6 : : 6 is to what?	*Ans.* 9.
2 : 9 : : 8 is to what?	*Ans.* 36.
5 : 7 : : 15 is to what?	*Ans.* 21.

RULE.

Set that number which is of the name or kind in which the answer is required, in the *third place:*

* RATIO is the relation of one thing to another of the same kind in regard to magnitude or quantity.

And, if the answer must be *greater* than the third term, set the *greater* of the remaining two terms in the *second*, and the less in the first place; but, if the answer must be *less* than the third term, set the *less* in the second, and the greater in the first place.

When the *first* and *second* terms are not of the same denomination, reduce one or both of them till they are; and, if the third consist of several denominations, reduce it to the lowest, then

Multiply the *second* and *third* terms together and divide the product by the FIRST, and the quotient will be the fourth term or answer.

NOTE.—The answer will be of the same denomination as the *third term*; and, in many instances, must be reduced to a greater denomination.

EXAMPLES.

1. If four pounds of sugar cost 50 cents, what will 24 pounds cost at the same rate? Ans. $3,00.

1st Term.		*2d Term.*		*3d Term.*
lbs.		*lbs.*		*cts.*
As 4	:	24	: :	50
		50		
		4)1200		
		3,00		

In this question the answer is required to be money: therefore *money* (the 50 cts.) must be in the *third place*. Because 24 pounds will cost *more* than 4 pounds—the *greater* (24 lbs.) must occupy the *second place*: and the remaining term (4 lbs.) the *first*

2. If 24 pounds cost 300 cents (or 3 dollars;) how many pounds may be purchased for 50 cents at the same rate? Ans. 4 lbs.

cts.		*cts.*		*lbs.*
As 300	:	50	: :	24
				50
				300)1200
				4

In this question the answer is required to be in *pounds*; therefore *pounds* (the 24) must be in the *third place*.

Because 50 cts. will purchase *less* than 300 cts.—the *less* (50 cts.) must occupy the *second place*: and the remaining term (300 cts.) the *first*.

3. Bought a load of corn containing 27 bushels, 3 pecks, at 50 cts. per bushel, what did it cost?
Ans. $13,87½.

```
bu.      bu. pe.      cts.
 1   :   27   3  : :  50
 4        4
---      ----
 4       111
          50
        ----
      4)5550
        ----
      $13,87½
```

Because pecks occur in the second term, the first and second are reduced to pecks.

4. What are 42 gallons worth, if 3 gallons 2 quarts cost $1,20? Ans. $14,40.

```
gals. qts.   gals.     D. cts.
 3    2   :   42   : :  1,20
 4             4
---          ----
14           168
              120
            -----
         14)20160
            -----
           $14,40
```

As quarts occur in the first term, the first and second are reduced to quarts.

5. If 8 bushels 2 pecks cost $4,25, how many bushels can I purchase with $38,25? Ans. 76 bu. 2 pe.

```
D. cts.      D. cts.     bu. pe.
4,25    :    38,25   : :  8   2
                34        4
             -----       --
             15300       34 pe.
            11475
            ------
     425)130050(306 pe.
          1275
          ----           pe.
            2550      4)306
            2550        ---
                     76 bu. 2 pe.
```

As two denominations occur in the third term, it is reduced to the less; hence the result is pecks, which must be reduced to bushels.

6. What will 5 lb. 6 oz. 5 dwt. of silver-ware cost at $1,50 per ounce? Ans. $99,37½.

```
oz.     lb. oz. dwt.      D. cts.
1   :    5  6  5   : :    1,50
20      12
--      --
20      66
         20
       ----
       1325
         150
       -----
       66250
      1325
      ------
   20)198750
      ------
      $99,37½
```

As dwts. are in the second term, the first and second must be reduced to dwts.

7. When 3 yards and 8 feet of plastering cost $1,40, what will be the cost of 16 yards? Ans. $5,76.

```
yds. ft.       yds.        D. cts.
 3   8    :    16    : :   1,40
 9              9
--            ---
35            144
               140
             -----
              5760
             144
             -----
         35)20160(5,76 cts.
```

8. How many yards of cloth can be purchased for 95 dollars, if 4 yd. 3 qr. cost $9,50?

Ans. 47 yd. 2 qr.; or 47½ yd.

```
    $ ct.     $ ct.      yd. qr.     qr.
As 9,50  :  95,00   : :   4   3    4)190(47 yd. 2 qr.
```

NOTE.—The operation may, in many instances, be contracted by dividing the second or third term by the first; or the first by either of the others, or by any number that will divide the first and either of the others without a remainder; and, using the quotients instead of the original numbers.

9. If 24 yards cost $96, what will 8 yards cost? Ans. $32.

yds.		*yds.*		*D.*		*yds.*		*yds.*		*D.*
24*a*	:	8*a*	: :	96*c*	*or*	24*a*	:	8	: :	96*a*
								4		
3*c*				Ans. 32.						4
								32		

10 If 36 bushels cost $72; what will 12 bu. cost? Ans. $24.

bu.		*bu.*		*D.*	*bu.*		*bu.*		*D.*
36*a*	:	12*a*	: :	72*c*	12)36	:	12	: :	72*a*
3*c*				24	3*a*		1		24

APPLICATION.

1 When 4 bushels of apples cost $2,25, what must be paid for 20 bushels? Ans. $11,25.

bu.		*bu.*		*D. cts.*
4	:	20	: :	2,25

2 How many yards of cloth can I buy for $60, when 5 yards cost $12? Ans. 25 yds.

D.		*D.*		*yds.*
12	:	60	: :	5

3 If 6 horses eat 21 bushels of oats in a given time; how much will 20 horses eat in the same time? Ans. 70 bu.

4 If 20 horses eat 70 bushels of oats in a certain time; how much will 4 horses eat in the same time? Ans. 14 bu.

5 If a family of ten persons use 7 bushels 3 pecks of wheat in a month; how much will serve them when there are 30 in the family? Ans. 23 bu. 1 pe.

6 If 14 lbs. of sugar cost 75 cents, how many pounds can be bought for three dollars? Ans. 56 lbs.

7 If 4 hats cost 12 dollars, what will 27 hats cost at the same rate? Ans. $81.

8 If 20 yards of cloth cost $85, what will 324 yards cost at the same rate?

Ans. $1377.

9 If 2 gallons of molasses cost 70 cents, what will 2 hogsheads cost? Ans. $44,10.

10 If 1 yard of cloth cost $3,25 cts., what will be the cost of 6 pieces, each containing 12 yds. 2 qrs.?

Ans. $243,75 cts.

11 If 3 paces or common steps of a person be equal to 2 yards, how many yards will 160 paces make?

Ans. 106 yds. 2 ft.

12 If a person can count 300 in 2 minutes, how many can he count in a day? Ans. 216000.

13 What quantity of wine at 60 cts. per gallon can be bought for $37,80 cts. Ans. 63 gal.

14 If 8 persons drink a barrel of cider in 10 days, how many persons would it require to drink a barrel in 4 days?

Ans. 20.

15 If 8 yards of cloth cost $12, what will 32 yards cost? Ans. $48.

16 If 3 bushels of corn cost $1,20, what will 13 bushels cost? Ans. $5,20.

17 If 9 dollars will buy 6 yards of cloth, how many yards will 30 dollars buy? Ans. 20.

18 If a man drink 3 gills of spirits in a day, how much will he drink in a year? Ans. 34 gal. 1 pt. 3 gi.

19 If 12 horses eat 30 bushels of oats in a week, how many bushels will serve 44 horses the same time?

Ans. 110.

20 If a perpendicular staff 6 feet long, cast a shadow 5 feet 4 inches, how high is that tree whose shadow is 104 feet long at the same time? Ans. 117 feet.

EXERCISES.

1 If 12 acres, 2 roods, produce 525 bushels of corn, how many bushels will 62 acres, 2 roods produce?

Ans. 2625 bu.

2 If 7 men plough 6 acres, 3 roods in a certain time, how many acres will 96 men plough in the same time?

Ans. 92 A. 2 R. 11 Per. 12 yd.+

3 Suppose 3 men lay 9 squares* of flooring in 2 days; how many men must be employed to lay 45 squares in the same time? Ans. 15 men.

4 If 7 pavers lay 210 yards of pavement in one day; how many pavers would be required to lay 120 yards in the same time? Ans. 4 pavers.

5 If 2 hands saw 360 square feet of oak timber in 2 days; how many feet will 8 hands saw in the same time? Ans. 1440 feet.

6 An engineer having raised a certain work one hundred yards in 24 days, with 5 men; how many men must be employed to perform a like quantity in 15 days? Ans. 8 men.

7 If 3 paces or common steps be equal to 2 yards; how many yards will 160 such paces make? Ans. 106 yd. 2 ft.

8 If a carriage wheel in turning twice round, advance 33 feet 10 inches; how far would it go in turning round 63360 times? Ans. 203 miles.

9 Sound flies at the rate of 1142 feet in 1 second of time; how far off may the report of a gun be heard in 1 minute and 3 seconds? Ans. 13 miles, 5 fur., 0 poles, 2 yd.

10 If a carter haul 100 bushels of coal at every 3 loads; how many days will it require for him to load a boat with 3600 bushels, suppose he haul 9 loads a day? Ans. 12 days.

	bu.		*bu.*		*da.*		*da.*
As	300	:	3600	: :	1	:	12.

11. If 8 men can reap a field of wheat in 4 days; how many days will it require for 16 men to do it? Ans. 2 days.

12 Sold 10 yards of linen at 5 dollars 50 cents; what was it a yard? Ans. 55 cents.

---

*A SQUARE is 10 feet long and 10 feet wide, or 100 square feet. This measure is employed in estimating the quantity of flooring, roofing, weather-boarding, &c.

13 If 7 pounds of cheese cost 87½ cents; what must I pay for 122 pounds? Ans. 15 dol. 25 ct.

14 If 1 ounce of silver cost 72 cents; what will 3 pounds 5 ounces come to? Ans. 29 dol. 52 ct.

Why do you multiply the second and third terms together and divide by the first?

What will 24 pounds of bacon cost at 50 ct. for every 4 pounds?

	lb.		*lb.*		*ct.*		*ct.*
As	4	:	24	::	50	:	300

cts.
4) 50
12½ the price of 1 lb.
24
50
25
300 ct. the price of 24 lb.

If 4 pounds cost 50 cents, divide the 50 cents by 4, gives the price of 1 pound: thus 4 into 50 12½. If 1 pound cost 12½ cents, 2 pounds will cost twice that; three pounds three times, and 24 pounds will cost 24 times 12½ ct.; that is 300 cts. or 3 dol.

If the second and third terms be multiplied together, and their product divided by the first, the result will be the same as it is when the third is divided by the first, and the quotient multiplied by the second.

15 If 15 yards of broad cloth cost 80 dollars; what will 75 cost. Ans. 400 dollars.

16 A man bought 1½ yards linen for $2 50 cts.; what is the worth of 1 qr. 2 na. at the same rate. Ans. 62½ ct.

17 If 321 bushels of salt cost $240 75 cents; what was it per bushel? Ans. 75 cents.

18 If the moon move 13 deg. 10 min. 35 sec. in one day; in what time does it perform one revolution? Ans. 27 da. 7 hrs. 43 min.

	deg.	*min.*	*sec.*		*deg.*		*da.*
As	13	10	35	:	360	::	1

19 If a staff 4 feet long, cast a shade 7 feet on level ground; how high is a steeple whose shade is at the same time 198 feet. Ans. $113\frac{1}{7}$

20 If a man's annual income be 1333 dollars, and he spend 2 dollars 14 cents a day; what will he save at the end of one year? Ans. $551 90 ct.

21 Suppose A. owes B. 791 dols. 60 ct., and can pay only 37½ cts. on the dollar; how much must B. receive? Ans. $296 85 ct.

22 Bought 3 casks of raisins, each containing 3 cwt. 1 qr. 14 lb.; how much did they cost at $6 21 ct. per cwt.? Ans. $62,87½

PROPORTION—DIRECT AND INVERSE.

Hitherto, proportion has been treated in general terms; it now remains to consider the two kinds, DIRECT and INVERSE.

DIRECT PROPORTION is that in which *more* requires *more*, or LESS requires LESS. Thus:

yd. *yd.* *dol.*
As 2 : 124 : : 4

If 2 yd. cost 4 dol., 124 yd. being *more* than 2 yd., will cost *more* than 4 dol.

yd. *yd.* *dol.*
As 124 : 2 : : 248

And, if 124 yd. cost 248 dol.; 2 yd. being *less* will cost *less*.

That is, *more* yards require *more* money, and *less* yards cost *less* money

INVERSE PROPORTION is that in which *more* requires *less;* and *less* requires *more*. Thus:

If 12 men built a wall in 4 days; how many men can do it in 8 days?

It is supposed that 12 men perform a piece of work in 4 days: a like piece of work is to be done in 8 days; this will require a *less* number of men: that is, *more* days require *less* men.

STATED.

da. *da.* *m.* *m.*
As 4 : 8 inversely : : 12 . 6

OR:

da. *da.* *m.* *m.*
As 8 : 4 directly : : 12 . 6

Here it is supposed that 12 men performed a piece of work in 8 days: a like piece is to be done in 4 days; this will require *more* men. That is, *more* days require *less* men, and *less* days require *more* men.

All the past exercises in proportion are *Direct*—the following will be

INVERSE PROPORTION.

Questions in *Inverse Proportion*, may be stated and solved by the same rule that is given for Direct Proportion.

EXERCISES.

1 If 6 mowers mow a meadow in 12 days; in what time will 24 mowers do it? Ans. 3 da.

2 If a man perform a journey in 6 days, when the days are 8 hours long; in what time will he do it when they are 12 hours long? Ans. 4 da.

3 If, when wheat is 83 cents a bushel, the cent loaf weighs 9 ounces; what ought it to weigh when wheat is $1 24½ cts. a bushel? Ans. 6 oz.

4 If 100 dollars principal in 12 months gain 6 dollars interest; what principal will gain the same interest in 8 months? Ans. $150.

5. If 12 inches long and 12 inches wide, make 1 square foot; how long must a board be that is 9 inches wide, to make 12 square feet? Ans. 16 ft.

6 A. lent B. 500 dollars for 6 months; how long must B. lend A. 220 dollars to be equivalent?

Ans. 13 months, 19 days.+

7 There is a cistern having a pipe that will empty it in 12 hours; how many pipes of the same capacity will empty it in 15 minutes? Ans. 48 pipes.

8 A certain building was raised in 8 months by 120 workmen, but the same being demolished, it is required to be rebuilt in 2 months; how many workmen must be employed? Ans. 480 men.

9 If for 48 dollars 225 cwt. be carried 512 miles; how many hundred weight may be carried 64 miles for the same money? Ans. 1800 cwt.

*A month is estimated at 30 days, unless a particular month is referred to.

10 If 48 men can build a wall in 24 days; how many men can do it in 192 days? Ans. 6 men.

11 How many yards of carpeting 2 ft. 6 in breadth, will cover a floor that is 27 feet long and 20 feet wide? Ans. 72.

12 What quantity of shalloon that is 3 quarters wide, will line 7½ yards of cloth that is 1½ yd. wide? Ans. 15 yd.

13 How many yards of matting that is 3 quarters wide, will cover a floor that is 18 feet wide and 60 feet long? Ans. 160.

14 In what time will $600 gain the same interest that $80 will gain in 15 years? Ans. 2 years.

Questions.

What is proportion?

When is the proportion *direct?*

When is it *inverse?*

Why is the proportion *inverse* in the last question?

A. because it is *more* money requiring *less* time.

Why is the proportion inverse in the 11th question?

A. because the shalloon is *narrower* than the cloth; that is *less* width requiring *more* length.

Why is the 10th question *inverse?*

PROMISCUOUS EXERCISES.

1 A certain steeple standing upon level ground, casts a shadow to the distance of 633 feet 4 inches, when a staff 3 feet long, perpendicularly erected, casts a shadow of 6 feet 4 inches; what is the height of the steeple? Ans. 300 ft.

2 A ship's company of 15 persons is supposed to have bread enough for a voyage, allowing each person 8 ounces a day, when they take up a crew of 5 persons, with whom they are willing to share; what will be the daily allowance of each person now? Ans. 6 oz.

3 Bought 215 yards of broad cloth at 6 dollars a yard;

what was the prime cost, and how must I sell it per yard to gain $135 on the whole.

Ans. prime cost $1290,00; to be sold for $6,62½ per yard.

4 If 100 men can complete a piece of work in 12 days; how many can do it in 3 days? Ans. 400 men.

5 If a board be 4½ inches wide; how long a piece will it take to make 1 square foot? Ans. 32 in.

6 A pole, whose height is 25 feet, at noon casts a shadow to the distance of 33 feet 10 inches; what is the breadth of a river which runs due East at the bottom of a tower 250 feet high, whose shadow extends just to the opposite edge of the water? Ans. 338 ft. 4 in.

7 A plain of a certain extent having supplied a body of 3000 horses with forage for 18 days; how long would it have supplied 2000 horses? Ans. 27 da.

8 A piece of ground 1 rod wide and 160 rods long, makes 1 acre; how wide a piece must I have across the end of a farm 32 rods wide to make an acre?

Ans. 5 rods.

9 I have a floor 24 feet long, and 15 feet wide, which I wish to cover with carpeting that is 3 quarters of a yard wide; how many yards must I buy.

Ans. 53 yards, 1 foot.

10 How much land at $2,50 an acre must be given in exchange for 360 acres worth $3,75 an acre?

Ans. 540 acres.

11 What is the weight of a pea to a steel-yard, which is 39 inches from the centre of motion, will balance a weight of 208 lbs., suspended at the draught end 3 quarters of an inch? Ans. 4 lb.

12 If $28 will pay for the carriage of 6 cwt. 150 miles; how far should 24 cwt. be carried for the same money? Ans. 37½ miles.

COMPOUND PROPORTION;
OR
THE DOUBLE RULE OF THREE.

DIRECT AND INVERSE.

COMPOUND PROPORTION is two or more series of proportionals combined.

Five, seven, nine, or other odd number of terms, is always given to find a sixth, eighth, or tenth, &c., or answer.

Rule for the Statement.

Place the numbers that is of the denomination in which the answer is required to be, in the *third place*. Then:

Consider separately each pair of similar terms and place them agreeably to the rule for SIMPLE PROPORTION.

OR,

Work by two separate statements in simple proportion.*

Rule for the Solution.

Reduce the several pairs of terms to similar denominations as in single proportion, and the last to the lowest denomination given: Then

Multiply the two *initials*, or left hand terms together for a DIVISOR, and the other three for a DIVIDEND.

Divide the *latter* by the *former*, and the quotient will be the answer, in that denomination to which the *third term* was reduced.

EXAMPLES.

1. If 6 men in 8 days build 40 rods of wall, how much will 18 men build in 20 days? Ans. 300 rods.

---

* It would be well for the pupil to work each sum both ways.

```
    Men   men
As 6  :  18        rods
   da.   da.  : :  40
As 8  :  20
   --    ---
   48    360
          40
        -----
   48)14400(300
       144
       ---
        00
```

The answer is required to be given in rods: then *rods* must be the *third* term. If 6 men build 40 rods, 18 men will build *more*; then more (18 men) must occupy the *second*, and *less* (6 men) the first place.

If 8 days produce 40 rods, 20 days will produce *more;* then *more* (20 days) must occupy the *second*, and *less* (8 days) the *first* place.

N. B. The first pair, or two upper terms must be alike. Also the lower pair must be alike.—That is, both must be *men* or both *days*, both hours or both bushels, &c.

2. If 6 men in eight days eat 10lb. of bread, how much will 12 men eat in 24 days? Ans 60.

```
men  6 : 12}
days 8 : 24}  : : 10 lb.
         --
        288
         10
        ----
   48)2880(60 Ans.
      288
      ---
        0
```

Contracted.

```
6 : 12 2}
8 : 24 3}  : : 10lb.
       --
        6
       10
       --
       60 Ans.
```

3. Suppose 4 men in 12 days mow 48 acres, how many acres can 8 men mow in 16 days? Ans. 128A.

4. If 10 bushels of oats be sufficient for 18 horses 20 days, how many bushels will serve 60 horses 36 days, at that rate? Ans. 60bu.

5. Suppose the wages of six persons for 21 weeks be 288 dollars, what must 14 persons receive for 46 weeks? Ans. $1472.

6. If the carriage of 8cwt. 128 miles cost $12.80, what must be paid for the carriage of 4cwt. 32 miles? Ans. $1.60.

7. If 37lb. of beef be sufficient for 12 persons 4 days, how many lb. will suffice 38 men 16 days? Ans. 468lb. $10\frac{2}{3}$ oz.

8. If a man can travel 305 miles in 30 days, when the days are 14 hours long, in how many days can he travel 1056 miles, when the days are $12\frac{1}{2}$ hours long?

Ans. 116 days.+2540.

9. If the carriage of 24cwt. for 45 miles be 18 dollars, how much will it cost to convey 76cwt. 121 miles?

Ans. $153 26 cts.+720.

10. A person having engaged to remove 8000cwt. in 9 days; removed 4500cwt. in 6 days, with 18 horses: how many horses will be required to remove the balance in the remaining 3 days? Ans. 28 horses.

11. If 3 men reap 12 acres 3 roods in 4 days 3 hours, how many acres can 9 men reap in 17 days?

Ans. 153 acres.

men	men			
3 :	9		a.	r.
d. h.	d. :	:	12	3
4 3 :	17		4	
12*	12		—	
—	—		51	
51	204			
3	9			
—	—			
153	1836			
	51			
	—			
	1836			
	9180			
	—			

153)93636(612 roods, or 153 acres
918
—
183
153
—
306
306
—
. . .

Analysis.

If 3 m. reap 12 a. 3 r.
1 m. rea 4 a. 1 r. and
9 m. reap 38 a. 1 r.

If 4 d. 3h. reap 38 a.
1 r. 1 d. reap 9 a. and
17 d. reap 153 a. Ans.

*The day is here estimated at twelve hours.

12. If 40 men build 32 rods of wall in 8 days, working 10 hours each day; in how many days will 60 men build 48 rods, working 12 hours a day?

Ans. 6 days, 8 hours.

Mon	mon	
60 :	40	
rods	rods	days
32 :	48	: : 8
hours	hours	
12 :	10	

Multiply all tho initial terms (or 60, 32, and 12) together for a *divisor:* and the other four for a *dividend.*

13. If 36 men dig a cellar 60 feet long, 24 feet wide, and 8 feet deep, in 16 days, working 16 hours per day, how many men can dig a cellar 80 feet long, 40 feet wide, and 12 feet deep, in 20 days, working 12 hours per day? Ans. 128.

Questions.

What is *compound proportion?*
How do you state questions in compound proportion?
Which terms do you multiply together for a *divisor?*
Which for a *dividend?*
What other method is there?

PRACTICE.

PRACTICE is a short and expeditious method of performing various calculations in business.

CASE 1.

When the given price is LESS *than one dollar.*

RULE.—Set down the given number as one dollar, and take such *aliquot part** or parts of that number, as the price is of one dollar, for the answer.

---

*An aliquot part of a number is any number that will divide it without a remainder; thus 4 is an aliquot part of 20; and 8 of 40; and 25 cents is an aliquot part of 100 cents, (or $1.) because 25 cts. are contained in 100 cts., an even number of times, without a remainder.

TABLE OF ALIQUOT PARTS.

CTS.			CWT.			ROODS.		
50	$\frac{1}{2}$	of a dollar	10	$\frac{1}{2}$	of a ton	roods		of an acre
$33\frac{1}{3}$	$\frac{1}{3}$		5	$\frac{1}{4}$		2	$\frac{1}{2}$	
25	$\frac{1}{4}$		4	$\frac{1}{5}$		1	$\frac{1}{4}$	
20	$\frac{1}{5}$		2	$\frac{1}{10}$				
$12\frac{1}{2}$	$\frac{1}{8}$		1	$\frac{1}{20}$		perches		
10	$\frac{1}{10}$		qr.		of a cwt.	20	$\frac{1}{8}$	
$6\frac{1}{4}$	$\frac{1}{16}$		2	$\frac{1}{2}$		10	$\frac{1}{16}$	
5	$\frac{1}{20}$		1	$\frac{1}{4}$		20	$\frac{1}{2}$	of a rood
4	$\frac{1}{25}$		lbs.			10	$\frac{1}{4}$	
2	$\frac{1}{50}$		16	$\frac{1}{7}$		8	$\frac{1}{5}$	
			14	$\frac{1}{8}$		5	$\frac{1}{8}$	
			8	$\frac{1}{14}$		4	$\frac{1}{10}$	
			7	$\frac{1}{16}$		2	$\frac{1}{20}$	

EXAMPLES.

1. What will 826 bushels of wheat come to at 25 cts. a bushel? Ans. \$206 50 cts.

cts.		\$
25	$\frac{1}{4}$	826
	\$	206,50

826 bushels, at one dollar a bushel, will cost 826 dollars: at 25 cents, or $\frac{1}{4}$ of a dollar, it will cost one fourth as much.

2. What will 934 gallons of molasses cost, at 50 cts. a gallon? Ans. \$467.

cts.		\$
50	$\frac{1}{2}$	934
	\$	467

3. What will 1832 bushels of salt cost, at 75 cents a bushel? Ans. \$1374.

cts.		\$
50	$\frac{1}{2}$	1832
25	$\frac{1}{2}$	
		916 cost at 50 c.
		458 cost at 25 c.
		\$1374 cost at 75 c.

At 50 cents the cost will be $\frac{1}{2}$ as much as—at one dollar.

At 25 cents the cost will be $\frac{1}{2}$ as much as—at 50 cts.

cts.		$
50	$\frac{1}{2}$	1832
25	$\frac{1}{4}$	
		916 ct. at 50.
		458 ct. at 25.
		$1374 ct. at 75.

As before; the cost at 50 cts. will be ½ as much as—at 1 dollar.

At 25 cts. it will be ¼ as much as—at 1 dollar.

4. What will 680 pounds of sugar cost, at 10 cents a pound? $68.

5. What will 742 pounds of pork cost, at 6¼ cts. a pound? Ans. $46 37½ cts.

6. What must I pay for 371 pounds of bacon, at 12½ cts. a pound? Ans. $46 37½ cts.

7. How much will 8750 bushels of rye cost, at 62½ cts. a bushel? Ans. $5468 75 cts.

8. How much must be paid for 4360 square feet of marble, at 87½ cents a foot? Ans. $3815.

9. What will 468 square yards of plastering cost, at 18¾ cents a yard? Ans. $87 75 cts.

10. How much will be the cost of laying 856 perches of stone, at 93¾ cents a perch? Ans. $802 50 cts.

11. What will the digging of a cellar, containing 180 cubic yards, cost, at 20 cents a yard? Ans. $36.

12. What will be the cost of hauling 248 cords of wood, at 31¼ cents a cord? Ans. $77 50 cts.

13. What must be paid for 432 perches of stone, at 37½ cts. a perch? Ans. $162.

14. How much must be given for 724 days labor, at 56¼ cents a day? Ans. $407 25.

15.	What will 742 bushels cost at	10 cts.	Ans. $74 20
16.	732	15	109 80
17.	732	20	146 40
18.	475	25	118 75
19.	684	30	205 20
20.	756	35	264 60
21.	927	40	370 80
22.	824	50	412 00
23.	682	55	375 10
24.	341	60	204 60
25.	784	70	548 80

28. What will	352 bushels cost at	6¼ cts.?	Ans. $22 00.
29.	436	12½	54 50
30.	724	18¾	135 75
31.	956	31¼	298 75
32.	742	37½	278 25
33.	274	43¾	119 87½
34.	732	56¼	411 75
35.	845	62½	528 12½
36.	684	68¾	470 25
37.	274	81¼	222 62½
38.	386	93¾	361 87½

Case 2.

When the given price is MORE *than one dollar.*

RULE.—Multiply the given sum by the number of dollars, and take the aliquot part or parts for the cents, as in Case 1.

EXAMPLES.

1. What will 342 cords of wood cost, at 3 dollars 75 cents a cord? Ans. $1282 50.

cts.		$	
50	½	342	
		3	
		1026	
25	½	171	
		85	50
		1282	50

342 cords at $1, will cost $342; at $3 it will cost 3 times $342; at 50 cts. it will cost ½ as much as it will at $1.; and at 25 cents, ½ as much as it will at 50 cts.; which added together, will be the cost at $3 75.

2. What will	250 acr. cost at	$4 62½	Ans. $1156 25
3.	435	5 87½	2555 62½
4.	273	6 12½	1672 12½
5.	942	7 37½	6947 25
6.	846	3 68¾	3119 62½
7.	957	5 75	5502 75
8.	236	6 93¾	1637 25
9.	754	3 56¼	2686 12½
10.	932	27 25	25397 00

Case 3.

When the given quantity consists of several denominations.

Rule.—Multiply the given price by the number of hundred weight, acres, yards, or pounds, &c. and take the aliquot parts for the quarters, roods, feet, or ounces, &c.

EXAMPLES.

1. What will 240 acres, 2 roods, 10 perches, cost at $15 25 cents an acre? Ans. $3668 57¾ cts.

```
                 $
2 r.  | ½ |    1525
      |   |     240
      |   |  ------
      |   |   61000
      |   |  3050
10 p. | ⅛ |    762½
      |   |     95¼+2 rem.
      |   |  ------
      |   |  366857¾
```

2. What will 29 yards, 4 feet, of stone pavement cost, at $2 25 cents a yard? Ans. $66 25 cts.

```
                     $
3 square feet | ⅓ | 225
              |   |  29
              |   | ----
              |   | 2025
              |   | 450
      1       | ⅓ |   75
              |   |   25
              |   | ----
              |   | 6625
```

3. What will 32 pounds 8 ounces of silver cost, at $15,62½ a pound? Ans. $510 41½.

6 oz. Troy	½	$15 62½
		32
		16
		3124
		4686
2 oz.	⅓	781¼
		260¼+2Rem.
		510 41½

4. What will 27 cwt. 3 qrs. cost, at $23 50 cts. a cwt.? Ans. $652 12½.

5. What will be the cost of 47 lb. 10 oz. (Troy) at $1 25 cts. Ans. $59 79.

6. What will 64 yds. 3 qrs. cost, at $2 25 a yard? Ans. $145 68¾.

7 Sold 83 yards 2 qrs. of cloth at $10 50 a yard; what does it amount to? Ans. $876 75.

8. What will the laying of 28 squares, 75 feet of flooring cost, at $2 25 cts. a square? Ans. $64 68¾.

9. What is the cost of 27 cords, 96 feet of fire wood, at $3 75 a cord. Ans. $104 06¼

10. What is the value of 428 gals. 3 qts. at $1 40 cts. a gallon? Ans. $600 25 cts.

11. What is the value of 765 gals. 3 qt. 1 pt. at $2 18¾ cents a gallon? Ans. $1675 34¾ cts.

12. What is the value of 5 hhds. 31½ gals. at $47 12 cts. a hogshead? Ans. $259 16 cts.

13. What is the value of 17 hhds. 15 gals. 3 qts. at $64 75 cts. per hogshead? Ans. $1116 93 cts. 7m.

14. What is the value of 120 bu. 2 pecks, at 35 cents a bushel? Ans. $42 17 cts. 5 m.

15. What is the value of 780 bu. 2 pecks, 2 qts. at $1 17 cts. a bushel? Ans. $913 25 cts.+

16. What is the value of 1354 bu. 1 peck, 5 qts. 1 pt. at 25 cts. a bushel? Ans. $338 60 cts. 5m.+

17. What is the value of 35 acres 2 roods 18 perches, at 54 dollars 35 cts. an acre? Ans. $1935 53 cts. 9m.

Questions

What is *practice?*

What is the rule for the solution of questions in practice?

What is an aliquot part?

Are 50 cts. an aliquot part of 100 cents?

What part of $1 is fifty cents?

What part of $1 is $33\frac{1}{3}$ cents?

What part of $1 is 25 cents?

What part of $1 is $12\frac{1}{2}$ cents?

What part of $1 is 10 cents?

What part of $1 is 20 cents?

What part of $1 is 5 cents?

What part of $1 is 4 cents?

What part of $1 is $6\frac{1}{4}$ cents?

TARE AND TRET.

TARE AND TRET are allowances made on the weight of some particular commodities.

Gross weight is the weight of the goods, together with the vessel that contains them.

Tare is an allowance for the weight of the vessel.

*Tret** is an allowance of 4 lb. for every 104, for waste, &c.

Neat weight is the weight of the goods, after all allowances are made.

RULE.

Subtract the *tare* from the *gross*, and the remainder is the *neat weight.*

EXAMPLES.

1 Bought a chest of tea, weighing gross 63 lb., tare 8 lb.—what are the neat weight and value, at 85 cents per lb?

*As tret is never regularly allowed in this country; no account of it is taken in this work.

lb.		*85 ct.*
63	gross—or, weight of the chest and tea	55 *lb.*
8	tare—or, weight of the chest	——
—		425
55	neat—or, weight of the tea itself	425
		——
		$46,75 value.

2 Bought 5 bags of coffee, weighing each 97 lb. gross, tare of the whole 7 lb.—what are the neat weight and value, at 25 cents per lb.? Ans. 478 lb. neat—$119,50.

3 The gross weight of a hogshead of sulphur is 1344 lb.; the tare 138 lb.—what are the neat weight and its value, at $4,75 per 100 lb.?

Ans. 1206 lb. neat—$57,28½.

4 Bought 3 barrels of sugar, weighing as follows, viz: 236 lb. gross, 23 lb. tare—217 lb. gross, 22 lb. tare—225 lb. gross, 23 lb. tare—what are the neat weight and value, at $8 per 100 lb.? Ans. 610 lb. neat—$48,80.

5 Sold 3 hogsheads of sugar, weighing each 12 cwt. 2 qrs. 14 lb. gross; tare 2 cwt. 1 qr. 27 lb.—what are the neat weight and value, at $11,50 per cwt.?

Ans. 35 cwt. 1 qr. 15 lb. neat—$406 91½ cts.

6 What is the neat weight of 15 tierces of rice, weighing 48 cwt. 3 qrs. 12 lb. gross; tare 6 cwt. 12 lb., and what is the value, at $5,25 per cwt.?

Ans. 42 cwt. 3 qrs. neat—$224,43¾.

7 What is the neat weight of 28 hogsheads of tobacco, weighing 201 cwt. 3 qrs. 12 lb. gross; tare 3140 lb.; and what does it come to at $5 per cwt.?

Ans. 173 cwt. 3 qrs. 8 lb—$869 10¾ cts.

8 Bought 17 bags of grain, weighing 3561 lb. gross; tare 2 lb. per bag—what is the neat? Ans. 3527 lb.

9 What is the neat weight of 16 bags of pepper, each weighing 65 lb. gross; tare 1½ lb. per bag—and what is the amount at 30 cents per lb.?

Ans. 1016 lb. neat—$304,80.

10 In 14 hogsheads of sugar, weighing 89 cwt. 3 qrs. 17 lb. gross; tare 100 lb. per hhd.—how much neat weight, and what is its value, at $9 per cwt.?

Ans. 77 cwt. 1 qr. 17 lb. neat—$696,61½.

11 What are the neat weight and value of 16 hhds. of tobacco, each weighing 5 cwt. 1 qr. 19 lb. gross; tare 101 lb. per hhd., at £2 0s. 10d. per cwt.?

Ans. 72 cwt. 1 qr. 4 lb. neat—£169. 5s. 4½d.

12 Bought 6 hhds. of sugar, each 1126 lb. gross; tare 117 lb. per hhd.—what are the neat weight and value at $8,75 per cwt.? Ans. 6054 lbs. $529,72½.

13 What are the neat weight and cost of a hogshead of sugar weighing gross 986 lb.; tare 12 per cent, (or 12 lb. for every 100 lb.) at $8 per neat hundred pounds?

lb. *lb.* *lb.* *lb.*		*lb.* *lb.* *lb.* *lb.*
As 100 : 986 : : 12 : 118,		Or as 100 : 88 : : 986 : 868,
lb.		*lb.*
986	gross.	868
118,	tare.	8 dol.
868,	neat weight.	$69,44 the value.

14 What are the neat weight and value of 4 hhds. of sugar weighing gross 4500lb. tare 12 lb. per cent. at $8, 75 per cent.? Ans. 3960 lbs. neat —$346,50.

15 Bought 10 hhds. of sugar, each 920 lb. gross; tare 10 lb. per cent.—what are the neat weight and value at $9,25 per cwt.? Ans. 8280 lb. neat– $765,90.

16 Sold 3 casks of alum, each 675 lb. gross; tare 13 lb. per cent.—what are the neat weight and value at $4, 25 per cent. Ans. 1762 lb. neat—$74,87.4375.

Or, 1762 lb. neat—$74,88.5nearly.

17 What is the neat weight of 4800lb.gross: tare 12 lb. per cent.? Ans. 4224 lb.

18 What are the neat weight and value of 4 hhds. of sugar, each 12 cwt. 1 qr. 14 lb. gross; tare 12 lb. per cwt. at $12,20 per cwt.?

Ans. 44 cwt. 22 lb. neat—$539 19½ cts.

19 Bought 17 hhds. of sugar, weighing 201 cwt. 2 qrs. 13 lb. gross; tare 10 lb. per cwt.—what are the neat weight and value at $14 per cwt.?

Ans. 183 cwt. 2 qrs. 13 lb. neat— $2570 62½ cts.

INTEREST.

INTEREST is an allowance made for the use of money.

Principal is the sum for which interest is to be computed.

Rate per cent. per annum is the interest of 100 dollars for one year.

Amount is the principal and interest added together.

CASE 1.

When the time is one year and the rate per cent. is any number of dollars.

RULE.—Multiply the principal by the rate per cent., and divide by 100; the quotient will be the interest for one year.

EXAMPLES.

1. What is the interest of 500 dollars for 1 year, at 6 per cent. per annum?

$500
6

100÷$30|00 Ans.

2. What is the interest of 225 dollars for 1 year, at 7 dollars per cent. per annum? Ans. $15 75.

3. What is the interest of 384 dollars 50 cents, for 1 year, at 5 dollars per cent. per annum? Ans. $19 22½.

4. What is the *amount* of $275 for 1 year, at 6 per cent. per annum? Ans. $291 50.

$275
6

16,50 interest
275,00 principal

$291,50 amount

5. What is the interest of 1654 dollars 81 cents for 1 year, at 5 dollars per cent. per annum? Ans. $82 74+.

6. What is the interest of 1500 dollars, for 1 year, at ½ dollar per cent. per annum? Ans. $7 50.

7. What is the amount of $736, at 6 per cent. per annum, for 1 year. 780 dols. 16.

8. What is the interest of 524 dollars, for 1 year, at 5¼ dollars per cent. per annum? Ans. $27 51.

9. What would be the interest of 842 dollars, for 1 year, at 5½ dollars per cent. per annum? Ans. $46 31.

Case 2.

When the interest is required for several years.

Rule.—Find the interest for one year, and multiply the interest for one year by the number of years.

EXAMPLES.

1. What is the interest of 500 dollars, for 4 years, at 6 dollars per cent. per annum?

```
 $500
    6
 ----
30 00
    4
 ----
$120 00 Ans.
```

2. What will be the interest of 540 dollars, for 2 years, at 5 dollars per cent. per annum? Ans. $54 00.

3. What would be the interest of 482 dollars, for 7 years, at 6 dollars per cent. per annum? Ans. $202 44.

4 What is the amount of $736 81¼ with 7 years, nine months interest due on it, at 6 per cent. per annum? Ans. $1079 43¾.

Note.—If the interest is required for years and months, multiply the interest for 1 year by the number of years, and take the aliquot parts of the interest for 1 year, for the months.

```
                $736 81¼
                       6
6 mo. | ½ |  4420,87½  interest 1 year
                       7
3 mo. | ½ | 30946,12½  interest 7 years
             2210,43¾  interest 6 months
             1105,21⅞+          3
            34261,78  interest for 7 yr. 9 mo.
            73681,25  principal
          ——————$107943 03 amount
```

5. What is the amount of $362 25 for 4 years 6 mo. at 6 per cent. per annum? Ans. $460 05¾.

CASE 3.

When the interest is required for any number of months, weeks or days, less or more than one year.

RULE.—Find the interest of the given sum for one year Then, by proportion,

As 1 year
Is to the given time,
So is the interest of the given sum (for 1 year)
To the interest for the time required.

Or take the aliquot parts of the interest for one year, for the given time, as in note, Case 2.

EXAMPLES.

1. What is the interest of $560 for 2 years and 6 mo. at 5 per ct. per annum? Ans $70.

```
           |  560
           |    5
6 mo. | ¼ | 2800  interest for 1 year
           |    2  years
           | 5600
             1400
          $70 00  interest for 2 years 6 months.
```

2. What is the interest of 325 dollars, for 4 years and 2 months, at 4 dollars per cent. per annum?

Ans. $54 16 cts. 6m.

3. What is the interest of 840 dollars for 5 years and 3 months, at 4 dollars per cent. per annum?

Ans. $176 40.

4. What is the interest of 840 dollars, for 5 years and 4 months, at 7 dollars per cent. per annum?

Ans. $313 60.

5. What is the interest of 560 dollars, for 4 months, at 6 dollars per cent. per annum?

```
    560
      6
   ----         m.   m.     $ cts.   $ cts.
100)33 60    As 12 : 4 : : 33 60 : 11 20  Ans.
```

6. What is the interest of 1200 dollars, for 15 weeks, at 5 dollars per cent. per annum? Ans. $17 30.

7. What will be the interest of 240 dollars, for 61 days, at 4¾ dollars per cent. per annum? Ans. $1 90.+

8. What is the interest of $1000, for 14 months, at 7 per cent. per annum? Ans. $81 66⅔.

9. What is the interest of 450 dollars, for 6 months and 20 days, at 5½ dollars per cent. per annum?

Ans. $13 75.

10. What is the interest of 375 dollars 25 cents, for 3 years 2 months 3 weeks and 5 days, at 6 dollars per ct. per annum? Ans. $72 92.+

11. What is the amount of $736 for 28 weeks, at 10 per cent. per annum? Ans. $775 63.

Case 4.

To find the interest of any sum for any number of days, as computed at banks.

Rule.—Multiply the dollars by the number of days, and divide by 6; the quotient will be the answer in mills.

The interest of any number of dollars for 60 days, at 6 per cent. will be exactly the number of cents; and if any other rate per cent. is required, take aliquot parts, and add or subtract according as the rate per cent. is more or less than 6.

EXAMPLES.

1. What is the interest of 563 dollars, for 60 days, at 6 per cent. per annum—and likewise at 7 per ct. per an.?

Ans. $5,63 at 6 per cent.
$6,56.8 at 7 per cent.

```
           $563
             60
          -----
         6)33780
in. at }  -----
6 per  }   5630 mills,        1 | 1/6 | $5630
cent.  }                        |     |   938
                                |     |  ----
              Interest at 7 per cent. 6568 mills.
```

2. What is the interest of 854 dollars, for 30 days, at 6 per cent. per annum? Ans. $4 27.

3. What is the interest of 1100 dollars, for 48 days, at 6 per cent. per annum? Ans. $8 80.

4. What is the interest of 3459 dollars, for 75 days, at 6 per cent. per annum? Ans. $43 23 cts. 7 m.+

5. What is the interest of 1500 dollars, for 60 days, at 5 per cent. per annum? Ans. $12 50.

Case 5.

The amount, time, and rate per cent. given, to find the principal.

Rule.—Find the amount of 100 dollars for the time required, at the given rate per cent.

Then, by proportion, as the amount of 100 dollars for the time required, (at the given rate per cent.) is to the amount given, so is 100 dollars to the principal required.

EXAMPLES.

1. What principal, at interest for 8 years, at 5 per ct. per annum, will amount to 840 dollars? Ans. $600.

```
  5 dollars
  8 years
 --
 40 Int. of $100 for 8 yr.
100                          $     $     $     $
---                         140 : 840 :: 100 : 600
140 Amt. of $100 for 8 yr.
```

2. What principal, at interest, for 6 years, at 4 per cent. per annum, will amount to $1240. Ans. $1000

3. What principal, at interest for 5 years, at 6 per ct. per annum, will amount to 2470 dollars? Ans. $1900.

Case 6.

The principal, amount, and time given, to find the rate per cent.

Rule.—Find the interest for the whole time given, by subtracting the principal from the amount.

Then, as the principal is to 100 dollars, so is the interest of the principal for the given time, to the interest of 100 dollars for the same time.

Divide the interest last found by the time, and the quotient will be the rate per cent. per annum,

Or by compound proportion.

EXAMPLES.

1. At what rate per cent. per annum, will 600 dollars amount to 744 dollars, in 4 years? Ans. 6 per cent.

```
                           $     $     $    $
$744 amount          As 600 : 100 : : 144 : 24.
 600 principal       yr.  $
 ---                 4) 24 (6 rate per cent.
 144 interest
```

Or by compound proportion:

```
      $     $
As 600 : 100         $       $
   yr.   yr.   : : 144  :  6 rate per cent.
   4  :   1
```

2. At what rate per cent. per annum, will $1200 amount to $1476, in 5 years and 9 months?

Ans. 4 per cent.

3. If 834 dollars, at interest 2 years and 6 months, amount to 927 dollars 82½ cents, what was the rate per cent. per annum? Ans. 4½ per cent.

CASE 7.

To find the time, when the principal, amount, and rate per cent. are given.

RULE.—Divide the whole interest by the interest of the principal for one year, and the quotient will be the time required, or by proportion.

EPAMPLES.

1. In what time will 400 dollars amount to 520 dollars, at 5 per cent. per annum? Ans. 6 years.

```
 $         $
400       520
  5       400
----      ---        $      $       Y.  Y
20|00    20)120(6    20  :  120  ::  1 : 6 Ans.
```

2. In what time will £1600 amount to £2048, at 4 per cent. per annum? Ans. 7 years.

3. Suppose 1000 dollars, at 4½ per cent. per annum, amount to 1281 dollars 25 cents, how long was it at interest? Ans. 6Y. 3mo.

COMPOUND INTEREST.

Compound interest is that in which the interest for one year is added to the principal, and that amount is the principal for the second year; and so on for any number of years.

RULE.—Find the amount of the given sum for the first year by simple interest, which will be the principal for the second year; then find the amount of the principal for the second year for the principal for the third year; and so on for any number of years.

Subtract the first principal from the amount, and the remainder will be the compound interest required.

EXAMPLES.

1. What is the compound interest of 150 dollars for 5 years, at 4 per cent. per annum?

Ans. $32,49

```
$150             $150
   4                6      inst 1st year
 ---              ---
6|00 int. 1 yr.   156      amount 1st year
                  6,24     int. 2d year
                 -----
$156             162,24    amount 2d year
   4               6,48.9  int. 3d year
----             -------
6|24             168,72.9  amount 3d year
                   6,74.9  int. 4th year.
                 -------
162,24           175,47.8  amount 4th year
     4             7,01.9  int. 5th year
------           --------
6|48.96          182,49.7  amount 5th year
                 150,00.0  principal
                 --------
                  32,49.7  compound int. for 5 years.
```

2. What is the compound interest of 760 dollars, for 3 years, at 6 dollars per cent. per annum?

Ans. $145 17 cts. 2 m.+

3. What is the compound interest of $242 50 cts., for 4 years, at 6 per cent. per annum?

Ans. $63 65 cents.

4. What is the amount of 1300 dollars, for 3 years, at 5 dollars per cent. per annum, compound interest?

Ans. $1504 91 cts. 2 m.+

5. How much is the amount of 3127 dollars, for 4 years, at 4½ dollars per cent. per annum, compound interest? Ans. $3729 00cts. 5m.

Questions.

What is interest?

What is the principal?

What is the rate per cent.

What is the amount?

How do you proceed when the interest for several years is required?

What is to be noted if the interest is required for years and months?

When the interest is required for any number of weeks or days, less or more than one year, how do you perform the operation?

How do you proceed to find the interest, at 6 per cent. for any number of days, as computed at banks?

What is to be observed when the interest is at any other rate than 6 per cent.?

How do you proceed, when the principal, amount, and time are given, to find the rate per cent.?

How do you find the time, when the principal, amount, and rate per cent. are given?

What is compound interest?

How is compound interest computed?

PROMISCUOUS EXERCISES.

1. What is the interest of 620 dollars 25 cents for 5 years, at 5½ per cent. per annum?
Ans. \$170 56 8 m +

2 What is the interest of \$420, for 1 year, at 7 per cent. per annum? Ans. \$29 40.

3 What is the interest of 1450 dollars, for 60 days, at 6 per cent. per annum? Ans. \$14 50 cts.

4 What is the compound interest of \$626 25, for 3 years, at 5¼ per cent. per annum?
Ans. \$103 91.+

5 What is the interest of \$1659 for 3 weeks, at 4 per cent. per annum? Ans. \$3 82½.+

6 In what time will 500 dollars amount to 1000 dollars at 8 per cent. per annum, simple interest?
Ans. 12 years, 6 months.

7 What principal, at interest for 6 years and 6 months, at 2 per cent. per annum, will amount to 250 dollars?
Ans. \$221 23 cts. 9 m.

8 At what rate per cent. per annum, will \$300 amount to \$450, in 5 years? Ans. 10 per cent.

INSURANCE, COMMISSION, AND BROKAGE.

INSURANCE, Commission and Brokage, are allowances made to insurers, factors, and brokers, at such rate per cent. as may be agreed on between the parties.

RULE.

Proceed in the same manner as though you were required to find the interest of the given sum for one year.

EXAMPLES.

1 What is the commission on 625 dollars, at 4 dollars per cent?

$$\begin{array}{r} \$625 \\ 4 \\ \hline \end{array}$$

Ans. $25,00

2 What is the commission on $1320, at 5 per cent.? Ans. $66.

3 What is the commission on 3450 dollars, at 4½ dollars per cent.? Ans. $155,25.

4 The sales of certain goods amount to 1680 dollars: what sum is to be received for them, allowing 2¾ dollars per cent. for commission? Ans. $1633,80.

5 What is the insurance of $760, at 6½ per cent.? Ans. $49,40.

6 What is the insurance of 5630 dollars, at 7¾ dollars per cent.? Ans. $436 32 cts. 5 m.

7 A merchant sent a ship and cargo to sea, valued at 17654 dollars: what would be the amount of insurance, at 18¾ dollars per cent.? Ans. $3310 12½ cts.

8 What is the brokage on 2150 dollars at 2 per cent.? Ans. $43.

9 When a broker sells goods to the amount of 984 dollars 50 cents, what is his commission, at 1¼ dollar per cent.? Ans. $12 30½ cts.+

10 If a broker buys goods for me, amounting to 1650

dollars 75 cents, what sum must I pay him, allowing him 1½ per cent.? Ans. $24 76 cts. 1 m.+

Questions.

What are Insurance, Commission, and Brokage?

How do you proceed to find the Insurance, Commission, or Brokage?

In what does this rule differ from interest? It takes no account of *time.*

DISCOUNT.

DISCOUNT is an abatement of so much money from any sum to be received before it is due, as the remainder would gain, put to interest for the given time and rate per cent.

RULE.

Find the interest of 100 dollars for the given time at the given rate per cent.

Add the interest so found to 100 dollars, then by proportion,

As the amount of 100 dollars for the given time,
Is to the given sum,
So is 100 dollars,
To the present worth.

If the discount be required, subtract the present worth from the given sum, and the remainder will be the discount.

NOTE.—When discount is made without regard to time, it is found precisely like the interest for one year.

EXAMPLES.

1. What is the present worth of 420 dollars, due in 2 years, discount at 6 per cent. per annum? Ans. $375.

$
6
2
12
100
112

$ $ $ $
112 : 420 :: 100 . 375

2 What is the present worth of 850 dollars, due in 3 months, at 6 per cent. per annum?
Ans. $837 43¾ cts.+

3 What is the discount of 645 dollars, for 9 months, at 6 per cent. per annum? Ans. $27 77½ cts.

4 What is the present worth of 775 dollars 50 cents, due in 4 years, at 5 per cent. per annum?
Ans. $646,25.

5 What is the present worth of 580 dollars, due in 8 months, at 6 per cent. per annum? Ans. $557,69.+

6 What is the present worth of 954 dollars, due in 3 years, at 4½ per cent. per annum?
Ans. $840 52 cts. 8 m.+

7 What is the discount of 205 dollars, due in 15 months, at 7 per cent per annum?
Ans. $16 49 cts. 5 m.+

8 Bought goods amounting to 775 dollars, at 9 months' credit: how much ready money must be paid, allowing a discount of 5 per cent. per annum?
Ans. $746 98 cts. 7 m.

9 I owe A. to the value of 1005 dollars, to pay as follows: viz. 475 dollars in 10 months, and the remainder in 15 months; what is the present worth, allowing discount at 6 per cent. per annum?
Ans. $945 40 cts. 4 m.

10 What is the difference between the interest of 2260 dollars, at 6 per cent. per annum, for 5 years, and the discount of the same sum for the same time and rate per cent.? Ans. $156 46 cts. 2m.+

11 What is the discount of 520 dollars, at 5 per cent.?

$520
5

$26,00 Ans.

12 How much is the discount of $782, at 4 per cent.? Ans. $31, 28

13 What is the discount of 476 dollars, at 3 per cent.? Ans. $14,28.

14 Bought goods on credit, amounting to 1385 dollars: how much ready money must be paid for them, if a discount of 6 per cent. be allowed? Ans. $1301,90.

15 I hold A.'s note for 650 dollars; but I agree to allow him a discount of 4½ per cent. for present payment: what sum must I receive? Ans. $620,75.

Questions.

What is discount?

What is first to be done?

After having found the interest of 100 dollars, at the given time and rate per cent., what is next to be done?

After having added the interest so found to 100 dollars or pounds, by what rule do you work to find the discount?

When discount is made without regard to time, how is it found?

EQUATION OF PAYMENTS.

EQUATION is a method of reducing several stated times, at which money is payable, to one mean, or equated time, when the whole sum shall be paid.

RULE.

Multiply each payment by its time, and divide the sum of all the products by the whole debt, the quotient will be the equated time.

Proof.—The interest of the sum payable at the equated time, at any given rate, will equal the interest of the several payments for their respective times.

EXAMPLES.

1 C. owes D. 100 dollars, of which the sum of 50 dollars is to be paid at 2 months, and 50 at 4 months; but they agree to reduce them to one payment; when must the whole be paid? Ans. 3 months.

$$50\times2=100$$
$$50\times4=200$$

100)300(3 months

2 A merchant hath owing to him 300 dollars, to be paid as follows: 50 dollars at 2 months, 100 dollars at 5 months, and the rest at 8 months; and it is agreed to make one payment of the whole; when must that time be? Ans. 6 months.

3 F. owes H. 2400 dollars of which 480 dollars are to be paid present, 960 dollars at 5 months, and the rest at 10 months; but they agree to make one payment of the whole, and wish to know the time? Ans. 6 months.

4 K. is indebted to L. 460 dollars which is to be discharged at 4 several payments, that is ¼ at 2 months, ¼ at 4 months, ¼ at 6 months, and ¼ at 8 months; but they agreeing to make one payment of the whole, the equated time is therefore demanded? Ans. 5 months.

5 P. owes Q. 420 dollars, which will be due 6 months hence, but P. is willing to pay him 60 dollars now, provided he can have the rest forborn a longer time: it is agreed on; the time of forbearance therefore is required? Ans. 7 months.

6 A merchant bought goods to the amount of 2000 dollars and agreed to pay 400 dollars at the time of purchase, 800 dollars at 5 months, and the rest at 10 months; but it is agreed to make one payment of the whole; what is the mean or equated time? Ans. 6 months.

BARTER.

BARTER is the exchanging of one kind of goods for another, duly proportioning their values, &c.

RULE.

The questions that come under this head, may be done by the compound rules, the Rule of Three, or Practice, as may be most convenient.

EXAMPLES.

1 A country storekeeper bought 150 bushels of salt, at 56 cents per bushel; and is to pay for it in corn, at 33⅓ cents per bushel; how much corn will pay for the salt?

	ct.		*ct.*		*bu.*		*bu.*
As	33⅓	:	56	: :	150	:	252

OR

```
ct.
 56          33⅓)8400
150            3    3
-----        ----------
2800         1|00)252|00
 56               -------
-----              252  bushels of corn.
Cost of the salt. 8400 cts.
```

2 How much wheat, at 1 dollar 25 cents per bushel, will pay for 35 sheep, at 2 dollars 25 cents a piece?
Ans. 63 bush.

3 How much sugar, at 9 cents per lb. will pay for 1 dozen pair of shoes, at 1 dollar 75 cents per pair?
Ans. 233⅓ lbs.

4 How much tea, at 80 cents per lb. will pay for 560 lbs. of pork, at 5 cents per lb.? Ans. 35 lbs.

5 Bought 4 hats for 3 dollars 50 cents; 4 dollars; 4 dollars 50 cents; and 5 dollars—how much corn at 32 cents will pay for them? Ans. 53 bush. 4 qts.

6 A. has 420 bushels of corn, which he barters with B. for oats, and is to receive 4 bushels of oats for 3 of corn—how many bushels of oats must A receive?
Ans. 560 bush.

7 A boy bartered 735 pears for marbles, giving 5 pears for 2 marbles—how many marbles ought he to have received? Ans. 294 marbles.

8 A boy exchanges marbles for pears, and gives 2 marbles for 5 pears—how many pears should he receive for 294 marbles? Ans. 735 pears.

9 A farmer bartered 3 barrels of flour, at 5 dollars 25 cents per barrel, for sugar and coffee, to receive an equal quantity of each—how much of each must he receive, admitting the sugar to be valued at 9 cents per lb. and the coffee at 14 cents? Ans. 68½ lb. nearly.

10 A bartered 42 hats, at 1 dollar 25 cents per hat, with B. for 50 pair of shoes, at 1 dollar 12½ cents per pair—who must receive money, and how much?

Ans. B. $3,75.

11 Sold 75 barrels of herrings, at 2 dollars 75 cents per barrel, for which I am to receive 75 bushels of wheat, at 1 dollar 8 cents per bushel, and the residue in money—how much money must I receive?

Ans. 125 dolls. 25 cts.

12 Sold 35 yards of domestic, at 20 cents per yard, and am to receive the amount in apples, at 25 cents per bushel—how many bushels must I have?

Ans. 28 bush.

13 Gave 35 yards of domestic for 28 bushels of apples, at 25 cents per bushel—what was the domestic rated at per yard? Ans. 20 cts.

14 What is rice per lb. when 340 lb. are given for 4 yards of cloth, at 4 dollars 25 cents per yard?

Ans. 5 cts.

15 Gave in barter 65 lbs. of tea for 156 gallons of rum, at 33⅓ per gallon—what was the tea rated at?

Ans. 80 cts. per lb.

16 Q. has coffee worth 16 cents per pound, but in barter raised it to 18 cts.; B. has broad cloth worth 4 dollars 64 cents per yard—what must B. raise his cloth to, so as to make a fair barter with Q? Ans. $5,22.

17 B. had 45 hats, at 4 dollars per hat, for which A. gives him 81 dollars 25 cents in cash, and the rest in pork, at 5 cents per lb; how much pork will be required?
Ans. 1975 lb.

18 Two merchants barter; A. receives 20 cwt. of cheese, at 2 dollars 87 cents per cwt.; B. 8 pieces of linen, at 9 dollars 78 cents per piece; which of them must receive money, and how much? Ans. A. $20,84.

19 If 24 yards of cloth be given for 5 cwt. 1 qr. of tobacco, at 5 dollars 7 cents per hundred; what is the cloth rated at per yard? Ans. $1.109.

20 A. barters 40 yards of cloth, at 98 cents per yard, with B. for 28½ lbs. of tea, at 1 dollar 53 cents per lb.; which must pay balance, and how much?
Ans. A. $4,405.

21 A has 7½ cwt. of sugar, at 8 cents per lb., for which B. gave him 12½ cwt. of cheese, what was the cheese rated at per lb.? Ans. $. 048.

22 What quantity of sugar, at 8 cts. per lb. must be given in barter for 20 cwt. of tobacco, at 8 dollars per cwt.? Ans. 17 cwt. 3 qrs. 12 lb.

23 P. has coffee, which he barters with Q. at 11 cts. per lb. more than it cost him, against tea, which stands Q. in 1 dollar 33 cents the lb., but he puts it at 1 dollar 66 cents; query, the prime cost of the coffee?
Ans. $. 443+

LOSS AND GAIN.

By Loss and Gain, merchants and dealers compute their gains or losses.

Rule.

Work by the Compound Rules, by Proportion, or in Practice, as may be most convenient.

EXAMPLES.

1 Bought 1234 lbs. of coffee, at 12½ cts. per lb., and sold the whole for 160 dollars; did I lose or gain by it, and how much? Ans. gained $5,75.

2 Bought 120 dozen knives, at 2 dollars 50 cents per dozen, and sold them at 18¾ cents a piece; did I gain or lose, and how much? Ans. lost $30.

3 Bought 1234 yards of muslin, for 17½ cents, and sold it at 20 cents per yard; what was the gain? Ans. $30,85.

4 Bought 10 chests of tea, each 63 lbs. neat, for 600 dollars, and retailed it at 87½ cents per lb.; did I gain or lose, and how much? Ans. lost 48 dol. 75 ct.

5 Gave 285 dollars 25 cents for 4564 lbs. of bacon, and sold it for 365 dollars 12 cents; what was the gain per lb? Ans. $. 1¾ cts.

6 Bought 1234 yards of muslin, for 246 dollars 80 cents, and sold it for 215 dollars 95 cents; what did I lose per yard? Ans. $. 2½ cts.

7 Gave 25 cts. per bushel for corn, and sold it at 28 cents; what is the gain per cent.? Ans. 12 dolls. per 100 dolls.

8 Sold corn at 25 cts. per bushel, and 4 cts. loss; what was the loss per cent.? Ans. $13,79.

9 Bought 13 cwt. 25 lbs. of sugar, for 106 dollars, and sold it at 9½ cts. per lb.; what did I gain per cent.? Ans. 32 dolls. 73 cts.

10 Bought 126 gallons of wine for 150 dollars, and retailed it at 20 cts. per pint; what was the gain per cent.? Ans. 34 dolls. 40 cts.

11 Sold a quantity of goods, for 748 dollars 66 cents, and gained 10 per cent; what did I give for them.? Ans. 680 dols. 60 cts.

	dols.	*dols.*	*dols.*
100	110 :	100 :	: 748,66
10			100
110			110)74866,00($680,60

12 Sold goods to the amount of $1234, and gained at the rate of 20 per cent.; what was the prime cost? Ans. $1028,33⅓

13 Sold a quantity of goods, for $475, and at a loss of 12 per cent.; what did I give for them?

	dols.	*dols.*	*dols.*
100	88 :	100 :	: 475
12			100
88			88)47500(539,77+Ans.

14 Sold hats to the amount of $136, at 20 per cent. loss; what was the first cost? Ans. $170.

15 Laid out $755 in salt; how much must I sell it for, so as to gain 12 per cent.?

12
100

As 100::112 : 755 :: 845,60 Ans.

16 Bought 32 yards of mole skin for 128 dollars; what must I sell it for per yard, so as to gain 20 per cent.? Ans. 4 dols. 80 cts.+

17 Bought 17 yards of silk for 21 dollars; how much per yard must I retail it for, and gain 25 per cent.? Ans. 1 dol. 54 cts.+

18 Bought 64 yards of muslin for 13 dollars 50 cents, but proving a bad bargain, I am willing to lose 8 per cent; what must I sell it at per yard? Ans.19 cts.4 m.+

19 When hats are bought at 48 cents, and sold at 54 cents; what is the gain per cent.? Ans. 12½

20 If, when cloth is sold for 84 cents per yard, there is gained 10 per cent.; what will be the gain per cent. when it is sold for 1 dollar 2 cents per yard?
Ans. 33 dols. 68 cts.+

21 Bought a chest of tea, weighing 490 lbs. for $122 50ct. and sold it for $137 20 cents; what was the profit on each lb.? Ans. 3 cts.

22 Bought 12 pieces of white cloth, for 16 dollars 50 cents per piece; paid 2 dollars 87 cents a piece for dying; for how much must I sell them each, to gain 20 per cent.? Ans. 23 dols. 244.

23 If 28 pieces of stuff be purchased at 9 dollars 60 cents per piece, and 10 of them sold at 14 dollars 40 cents, and 8 at 12 dollars per piece; at what rate must the rest be disposed of, to gain 10 per cent. by the whole?
Ans. 5 dols. 568.

24 Sold a yard of cloth for 1 dollar 55 cents, by which was gained at the rate of 15 per cent.; but if it had been sold for 1 dollar 72 cents; what would have been the gain per cent.? Ans. 27 dols. 69+

25 If, when cloth is sold at $. 935 a yard, the gain is 10 dollars per cent.; what is the gain or loss per cent., when it is sold at 80 cents per yard?
Ans. 5 dollars 88+loss.

26 A draper bought 100 yards of broad cloth, for which he gave $56—I desire to know how he must sell it per yard, to gain $19 in the whole?
Ans. 75 ct. per yard.

27 A draper bought 100 yards of broad cloth for $56; I demand how he must sell it per yard, to gain $15 in laying out $100? Ans. 64 ct. 4 m.

28 Bought knives at 11 cents, and sold them at 12 cents; what will I gain by laying out 100 dollars in knives? Ans. 9 dols. 09+

29 Bought knives at 11 cents, and sold them at 12 cents; what did I gain by selling to the amount of 100 dollars? Ans. 8 dols. 333+

30 If by selling 1 lb. of pepper for 10½ cents, there are 2 cents lost; how much is the loss per cent.? Ans. 16 dols.

31 A merchant receives from Lisbon, 180 casks of raisins, which stands him in here 2 dollars 13 cents each, and by selling them at 3 dollars 68 cents per cwt., he gains 25 per cent.; required the weight of each cask, one with another? Ans. 81 lb.

FELLOWSHIP.

Fellowship is a method by which merchants and others adjust the division of property, loss, or gain, &c., in proportion to their several claims.

Case 1. Simple Fellowship.

When the claims are in proportion to the amount of stock, labor, &c., without regard to time.

Rule. (*By Proportion.*)

As the whole amount of stock or labor,
Is to each man's portion,
So is the whole property, loss, or gain,
To each man's share of it.

Proof.—The sum of all the shares must equal the whole gain, &c.

EXAMPLES.

1 Two men bought a stock of goods for 480 dollars, of which A. paid 320, and B. 160. They gained 128 dollars by the transaction; what was the share of each?

Ans. A. received 85 dols. 33⅓ cts. and B. 42 dollars 66⅔ cts.

		Proof
	$ $ $ $ *ct.*	
A's. stock $320	As 480 : 320 :: 128 : 85,33⅓	$85,33⅓
B's. stock 160		42,66⅔
——	$ $ $ $ *ct.*	——
Whole st'k 480	As 480 : 160 :: 128 : 42,66⅔	128,00

2 Three workmen having undertaken to do a piece of work for 275 dollars, agreed to divide their profits in proportion to the amount of labor each one performed. M. labored 50 days, N. 65 days, and O. 85 days: what was the share of each?

Ans. M. received 68 dols. 75 cts.; N. 89 dols. 37½ cts.; and O. 116 dols. 87½ cts.

3 A merchant being deceased, worth 1800 dollars, is found to owe the following sums: to A. 1200 dollars, to B. 500 dollars, to C. 700 dollars: how much is each to have in proportion to the debt?

Ans. A. 900 dols., B. 375 dols, and C. 525 dols.

4 Three drovers pay among them 60 dollars for pasture, into which they put 200 cattle, of which A. had 50, B. 80, and C. 70: I would know how much each had to pay? Ans. A. 15 dols., B. 24 dols., C. 21 dols.

5 A man failing, owes the following sums: to A. 120 dollars, to B. 250 dollars 75 cents, to C. 300 dollars, to D. 208 dollars 25 cents; and his whole effects were found to amount to but 650 dollars: what will each one receive in proportion to his demand?

Ans. A. $ 88.73.+ C. $221.84.+
B. $185.42.+ D. $153.99.+

6 A bankrupt is indebted to A. 500 dollars 37½ cents——to B. 228 dollars—to C. 1291 dollars 23 cents—to D. 709 dollars 40 cents; and his estate is worth 2046 dollars 75 cents: how much does he pay per cent., and what does each creditor receive?

Ans. He pays 75 per cent., and A. receives 375 dollars 27¾ cts.; B. 171 dols.; C. 968 dols. 42¾ cts.; and D. 532 dols. 5 cts.

7 If a man is indebted to A. 250 dollars 50 cents, to B. 500 dollars, to C. 349 dollars 50 cents, but when he comes to make a settlement, it is found he is worth but 960 dollars, how much will each one receive, if it be in proportion to their respective claims?

Ans. { A. $218 61 cts. 8 m.+
B. $436 36 cts. 3 m.+
C. $305 01 ct. 8 m.+ }

Case 2. Compound Fellowship.

When the respective stocks are considered with relation to *time*.

Rule. (*By Proportion.*)

Multiply each man's stock by its time; add the several products together; then:
As the sum of the products
Is to each particular product,
So is the whole gain or loss
To each man's share of the gain or loss.

EXAMPLES.

1 Three merchants traded together; A put in 120 dollars for 9 months, B. 100 dollars for 16 months, and C. 100 dollars for 14 months, and they gained 100 dollars; what is each man's share?

	$		mo.		
A's. stock	120	×	9	=	1080
B's. stock	100	×	16	=	1600
C's. stock	100	×	14	=	1400
				Sum	4080

	Sum.	*Prod.*	$		$	
As	4080 :	1080 ::	100	:	26,47+	A's. share.
As	4080 :	1600 ::	100	:	39,21½+	B's. share.
As	4080 :	1400 ::	100	:	34,31+	C's. share.

2 Three men traded together; L. put in 88 dollars for 3 months, M. 120 dollars for 4 months, and N. 300 dollars for 6 months; they gained 184 dollars: what will each man receive of the gain?

Ans. { L. $ 19 09 cts. 4 m.
M. $ 34 71 cts. 6 m.
N. $130 18 cts. 8 m.

3 Two merchants entered into partnership for 16 months: A. put in at first $600, and at the end of 9 months put in $100 more; B. put in at first $750, and at the end of 6 months took out $250, at the close of the time their gain was $386, what was the share of each?

Ans. A's. share was $200,79¾; B's. share was $185,20.

4 A., B., and C., made a stock for 12 months; A. put in at first $873,60, and 4 months after he put in $96,00 more; B. put in at first $979,20, and at the end of 7 months he took out $206,40; C. put in at first $355,20, and 3 months after he put in $206,40, and 5 months after that he put in $240,00 more. At the end of 12 months, their gain is found to be $3446,40; what is each man's share of the gain?

Ans. { A's. share is $1334,82½
B's. - - $1271,61½+
C's. - - $839,96 }

Questions.

What is Fellowship?
By what rule are its operations performed?
When is Fellowship simple?
When is it compound?
In what respect is Fellowship compound?

Ans. The proportion is compound: that is, the division of property, gain, &c., is founded on the compound proportion of the stock and time.

VULGAR FRACTIONS.

A VULGAR FRACTION is a part, or parts of a unit expressed by two numbers placed one above the other with a line between them. As $\frac{1}{2}$, $\frac{2}{3}$, &c.

The number below the line is the *denominator*, the number above the line is the *numerator*.

The *denominator* denotes the number of parts into which the *unit* is divided.

The *numerator* shows how many of those parts are to be taken.

Fractions are either *proper*, *improper*, or *compound*.

A *proper* fraction is one whose numerator is *less* than its denominator, as $\frac{5}{8}$ or $\frac{3}{7}$.

An *improper* fraction is one whose numerator is *greater* than its denominator, as $\frac{8}{5}$ or $\frac{9}{4}$.

A *compound* fraction is a fraction of a fraction, as $\frac{1}{2}$ of $\frac{2}{3}$, or $\frac{2}{3}$ of $\frac{3}{4}$.

A *mixed* number is a whole number and a fraction.

REDUCTION OF VULGAR FRACTIONS.

Case 1.

To reduce a fraction to its lowest terms.

RULE.

Divide the terms by any number that will divide both without a remainder, and divide the quotient in the same manner, and so on till no number greater than one will divide them: the fraction is then at its lowest terms.

EXAMPLES.

1. Reduce $\frac{84}{108}$ to its lowest terms.

   $4)\frac{84}{108}=\frac{21}{27}=\frac{7}{9}$ result.

2. Reduce $\frac{9}{36}$ to its lowest terms. Res. $\frac{1}{4}$
3. Reduce $\frac{10}{120}$ to its lowest terms. Res. $\frac{1}{12}$
4. Reduce $\frac{48}{56}$ to its lowest terms. Res. $\frac{6}{7}$

NOTE.—When a divisor cannot readily be found, divide the denominator by the numerator, and that divisor by the remainder, and so on, till nothing remain: the last divisor is the common measure of the two numbers; with which proceed as before.

5. Reduce $\frac{35}{136}$ to its lowest terms. Res. $\frac{5}{8}$

5 Reduce $\frac{85}{136}$ to its lowest terms. Res. $\frac{5}{8}$.

```
      85
     ----
  85)136(1
      85
     ----
     51)85(1
        51
        --
        34)51(1
           34
           --
           17)34(2
              34
```

Here 17 being the last divisor, is the common measure of 85 and 136.

$$17 \left| \frac{85}{136} \right. = \frac{5}{8}$$

6. Reduce $\frac{91}{117}$ to its lowest terms. Res. $\frac{7}{9}$.
7. Reduce $\frac{144}{216}$ to its lowest terms. Res. $\frac{2}{3}$.
8. Reduce $\frac{3619}{6251}$ to its lowest terms. Res. $\frac{11}{19}$.

Case 2.

To reduce a mixed number to an improper fraction.

RULE.

Multiply the whole number by the denominator, and add the numerator to the product for the numerator of the improper fraction, and place the denominator under it.

EXAMPLES.

1. Reduce 12 $\frac{4}{9}$ to an improper fraction. Res. $\frac{112}{9}$.

```
 12 4/9
  9
 ---
 112
 ---
  9
```

Nine 12's are 108; add 4 makes 112 ninths.

2. Reduce 17 $\frac{3}{7}$ to an improper fraction. Res. $\frac{122}{7}$.
3. Reduce 45 $\frac{2}{3}$ to an improper fraction. Res. $\frac{137}{3}$.
4. Reduce 24 $\frac{13}{17}$ to an improper fraction. Res. $\frac{421}{17}$.

Case 3.

To reduce an improper fraction to its proper value.

RULE.

Divide the numerator by the denominator, and the quotient will be the whole number; the remainder, if any, will be the numerator of the fraction.

EXAMPLES.

1 Reduce $\frac{17}{5}$ to its proper value. Res. $3\frac{2}{5}$.

$\frac{17}{5}$ 5)17

$3\frac{2}{5}$

2 Reduce $\frac{112}{9}$ to its proper terms. Res. $12\frac{4}{9}$.
3 Reduce $\frac{122}{7}$ to its proper terms. Res. $17\frac{3}{7}$.
4 Reduce $\frac{421}{17}$ to its proper terms. Res. $24\frac{13}{17}$.

Case 4.

To reduce several fractions to other fractions having a common denominator, and retaining their value.

RULE.

Multiply each numerator into all the denominators but its own, for the respective numerators; and all the denominators together, for a common denominator.

EXAMPLES.

1 Reduce $\frac{2}{3}$ $\frac{3}{4}$ and $\frac{5}{6}$ to a common denominator.
Res. $\frac{48}{72}$ $\frac{54}{72}$, and $\frac{60}{72}$.

$2\times4\times6=48$ }
$3\times3\times6=54$ } Numerators
$5\times3\times4=60$ }
$3\times4\times6=72$, common denominator.

Then we have $\frac{48}{72}$ for $\frac{2}{3}$; $\frac{54}{72}$ for $\frac{3}{4}$, and $\frac{60}{72}$ for $\frac{5}{6}$.

Reduce each new fraction to its lowest terms, and the result will prove the work to be right.

2. Reduce $\frac{3}{4}$, $\frac{5}{6}$, and $\frac{7}{12}$, to a common denominator
Res. $\frac{216}{288}$, $\frac{240}{288}$, and $\frac{168}{288}$.

3. Reduce $\frac{1}{2}$, $\frac{2}{3}$, $\frac{5}{6}$, and $\frac{7}{12}$, to a common denominator.
Res. $\frac{216}{432}$, $\frac{288}{432}$, $\frac{360}{432}$, and $\frac{252}{432}$.

4. Reduce $\frac{7}{8}$, $\frac{2}{3}$, $\frac{3}{5}$, and $\frac{5}{6}$, to a common denominator.
Res. $\frac{630}{720}$, $\frac{480}{720}$, $\frac{432}{720}$, and $\frac{600}{720}$.

NOTE.—It is often convenient to use the *least possible* common denominator; to find which, divide the denominators by any number that will divide two or more of them without a remainder, setting down those that would have remainders; then multiply all the divisors and all the quotients together.

4	2	3	4	5	6	7	8	9
3	2	3	1	5	6	7	2	9
2	2	1		5	2	7	2	3
	1			5	1	7	1	3

$4\times3\times2\times5\times7\times3=2520$ common denom. Which may be divided separately by 2, 3, 4, 5, 6, 7, 8, and 9, without a remainder.

EXAMPLES.

5 Find the *least* common denominator for $\frac{3}{4}$, $\frac{7}{8}$, $\frac{11}{12}$, $\frac{5}{16}$, and $\frac{9}{20}$, and compute their equivalent fractions.

Res. $\frac{180}{240}$, $\frac{210}{240}$, $\frac{220}{240}$, $\frac{75}{240}$, $\frac{108}{240}$.

240 com. denom.

$\frac{3}{4}$	$60\times3=180$
$\frac{7}{8}$	$30\times7=210$
$\frac{11}{12}$	$20\times11=220$
$\frac{5}{16}$	$15\times5=75$
$\frac{9}{20}$	$12\times9=108$

6 Reduce $\frac{3}{4}$, $\frac{5}{6}$, $\frac{7}{8}$, $\frac{9}{10}$, and $\frac{11}{12}$, to their least common denominator. Res. $\frac{90}{120}$, $\frac{100}{120}$, $\frac{105}{120}$, $\frac{108}{120}$, and $\frac{110}{120}$.

7 Reduce $\frac{5}{8}$, $\frac{7}{10}$, $\frac{9}{16}$, $\frac{11}{24}$, to their least common denomnator. Res. $\frac{150}{240}$, $\frac{168}{240}$, $\frac{135}{240}$, and $\frac{110}{240}$.

8 Reduce the above fractions to a common denominator, by the general rule, Case 4.

Res. $\frac{19200}{30720}$, $\frac{21504}{30720}$, $\frac{17280}{30720}$, $\frac{14080}{30720}$.

$5\times10\times16\times24=19200$
$7\times8\times16\times24=21504$
$9\times8\times10\times24=17280$
$11\times8\times10\times16=14080$

CASE 5.

To reduce a compound fraction to a simple one.

RULE.

Multiply the numerators together for a new numerator; and the denominators together for a new denominator.

EXAMPLES.

1 Reduce $\frac{3}{4}$ of $\frac{5}{6}$ of $\frac{9}{10}$ to a single fraction. Res. $\frac{9}{16}$.

$$\frac{3\times5\times9}{4\times6\times10}=\frac{135}{240}=\frac{9}{16}$$

2 Reduce $\frac{5}{9}$ of $\frac{4}{7}$ of $\frac{11}{12}$ to a single fraction. Res. $\frac{55}{189}$.
3 Reduce $\frac{2}{7}$ of $\frac{5}{9}$ of $\frac{4}{5}$ to a single fraction. Res. $\frac{8}{63}$.
4 Reduce $\frac{12}{14}$ of $\frac{5}{6}$ of $\frac{1}{2}$ to a single fraction. Res. $\frac{5}{14}$.

CASE 6.

To reduce a fraction of one denomination to the fraction of another denomination, but greater, retaining the same value.

RULE.

Multiply the denominator of the fraction by the number of that denomination which it takes to make one of the next, and so on to the denomination required, and place the numerator of the given fraction over it.

EXAMPLES.

1 Reduce $\frac{2}{3}$ of a quart to the fraction of a bushel.*

qt.

$$\frac{2}{3\times8\times4=96}\quad\frac{2}{96}=\frac{1}{48}$$

Result, $\frac{1}{48}$ of a bushel.

2 Reduce $\frac{3}{4}$ of an ounce, Troy, to the fraction of a pound. Res. $\frac{3}{48}$, or $\frac{1}{16}$ of a pound.
3 Reduce $\frac{4}{5}$ of a nail to the fraction of a yard. Res. $\frac{4}{80}$, or $\frac{1}{20}$ of a yard.
3 Reduce $\frac{7}{8}$ of a perch to the fraction of an acre. Res. $\frac{7}{1280}$.

* That is, what part of a bushel are two-thirds of a quart?

4 Reduce $\frac{9}{13}$ of a pint to the fraction of a hogshead
Res. $\frac{1}{728}$ of a hhd.

Case 7.

To reduce the fraction of one denomination to the fraction of another, but less, *retaining the same value.*

RULE.

Multiply the given *numerator* by the parts of that between it and that to which it is to be reduced, and place the product over the given denominator for the fraction required.

EXAMPLES.

1 Reduce $\frac{1}{48}$ of a bushel to the fraction of a quart.
Res. $\frac{2}{3}$ of a quart.
2 Reduce $\frac{1}{20}$ of a yard to the fraction of a nail.
Res. $\frac{4}{5}$ of a nail.
3 Reduce $\frac{7}{1280}$ of an acre to the fraction of a perch.
Res. $\frac{7}{8}$ of a perch.
4 Reduce $\frac{1}{728}$ of a hogshead to the fraction of a pint.
Res. $\frac{9}{13}$ of a pint.
5 Reduce $\frac{1}{1584}$ of a day to the fraction of a minute.
Res. $\frac{10}{11}$ of a minute.

Case 8.

To reduce a fraction to its proper value or quantity, in whole numbers.

RULE.

Multiply the numerator by the parts of the integer, and divide by the denominator.

EXAMPLES.

1 Reduce $\frac{7}{8}$ of a yard to its proper quantity.
Res. 3 qr. 2 na.

7 eighths of a yd.	4 eighths of a qr.
4	4
8)28	8)16
3 $\frac{4}{8}$ quarters	2 nails

2 Reduce $\frac{5}{9}$ of a pound, avoirdupois, to its proper quantity. Res. 8 oz. $14\frac{2}{9}$ dr.

3 Reduce $\frac{3}{4}$ of a pound, Troy, to its proper quantity. Res. 9 oz.

4 Reduce $\frac{4}{7}$ of a mile to its proper quantity. Res. 4 fur. 125 yd. 2 ft. $1\frac{5}{7}$ inch.

5 Reduce $\frac{7}{16}$ of an acre to its proper quantity. Res. 1 rood, 30 perch.

6 Reduce $\frac{3}{5}$ of a dollar to its proper quantity. Res. 60 cents.

7 Reduce $\frac{1}{3}$ of a pound to its proper value. Res. 6s. 8d.

8 Reduce $\frac{45}{73}$ of a year (365 days) to its proper quantity. Res. 225 days.

9 Reduce $\frac{7}{9}$ of a tun to its proper quantity. Res. 3 hhd. 7 gal.

10 Reduce $\frac{7}{9}$ of a ton to its proper quantity. Res. 15 cwt. 2 qr. 6 lb. 3 oz. $8\frac{8}{9}$ dr

Case 9.

To reduce a given quantity to a fraction of any greater denomination of the same kind.

RULE.

Reduce the given quantity to the lowest denomination mentioned for a numerator; and the integer to the same denomination, for a denominator.

EXAMPLES.

1 Reduce 3 qr. 2 na. to the fraction of a yard. Res. $\frac{7}{8}$ of a yard.

```
              qr.  na.
               3    2
               4
              ---
yd.            14)     (7
              ---} = {-
1×4×4=16)     (8
```

2 Reduce 2 roods 20 perches to the fraction of an acre. Res. $\frac{5}{8}$ of an acre.

3 Reduce 6 furlongs 16 poles to the fraction of a mile. Res. $\frac{4}{5}$ of a mile.

4 Reduce 9 ounces, Troy, to the fraction of a pound. Res. $\frac{3}{4}$ of a pound.

5 Reduce 7 hours 12 minutes to the fraction of a day. Res. $\frac{3}{10}$ of a day.

ADDITION OF VULGAR FRACTIONS.

RULE.—Reduce the given fractions, if necessary, to single ones, or to a common denominator; add all the numerators together, and place the sum over the common denominator.

EXAMPLES.

$\frac{5}{8}$	$4\frac{3}{10}$	$6\frac{2}{9}$
$\frac{1}{8}$	$3\frac{4}{10}$	$3\frac{5}{9}$
$\frac{3}{8}$	$7\frac{7}{10}$	$7\frac{4}{9}$
$\frac{7}{8}$	$3\frac{5}{10}$	$8\frac{7}{9}$
$\frac{3}{8}$	$7\frac{9}{10}$	$2\frac{8}{9}$
$\frac{19}{8}$ or $2\frac{2}{3}$	$26\frac{8}{10}=26\frac{4}{5}$	$28\frac{8}{9}$

NOTE 1.—When the fractions are of different denominators, reduce them to a common denominator, and proceed as above. (See *Note*, page 139.)

4 Add $\frac{3}{4}$, $\frac{7}{8}$, $\frac{11}{12}$, $\frac{5}{16}$ and $\frac{9}{20}$ together. Result $3\frac{73}{240}$.

	240
$\frac{3}{4}$	$60\times 3=180$
$\frac{7}{8}$	$30\times 7=210$
$\frac{11}{12}$	$20\times 11=220$
$\frac{5}{16}$	$15\times 5=75$
$\frac{9}{20}$	$12\times 9=108$

$$\frac{793}{240}=3\frac{73}{240}$$

5 Add $\frac{3}{4}$, $\frac{5}{6}$, $\frac{7}{8}$, $\frac{9}{10}$ and $\frac{11}{12}$ together. Res. $4\frac{11}{40}$.

6 Add $\frac{5}{9}$, $\frac{7}{10}$, $\frac{9}{16}$, and $\frac{11}{24}$ together. Res. $2\frac{83}{240}$.

7 Add $\frac{1}{8}$, $\frac{3}{10}$, and $\frac{4}{12}$ together. Res. $\frac{728}{960}=\frac{91}{120}$.

8 Add $\frac{2}{3}$ and $\frac{5}{7}$ together. Res. $1\frac{8}{21}$.

9 Add $\frac{7}{10}$, $\frac{11}{12}$ and $\frac{4}{9}$ together Res. $2\frac{11}{180}$.

NOTE 2.—When mixed numbers occur, place them as in examples 2 and 3; proceed with the fractions as directed in *Note* 1; and if they amount to one or more integers, carry them to the integers, and proceed as in simple addition.

10 Add $5\frac{2}{3}$, $6\frac{7}{8}$, and $4\frac{1}{2}$ together. Res. $17\frac{1}{24}$.

	24
$5\frac{2}{3}$	$8\times2=16$
$6\frac{7}{8}$	$3\times7=21$
$4\frac{1}{2}$	$12\times1=12$
$17\frac{1}{24}$	

$$\frac{49}{24}=2\frac{1}{24}.$$

11 Add $2\frac{1}{2}$ and $3\frac{3}{4}$ together. Resuit $6\frac{1}{4}$.
12 Add $7\frac{3}{7}$ and $5\frac{4}{9}$ together. Res. $12\frac{55}{63}$
13 Add $17\frac{1}{2}$ and $\frac{2}{3}$ together. Res. $18\frac{1}{6}$.
14 Add 4, 6, 9, $\frac{3}{4}$ and $\frac{5}{9}$ together. Res. $20\frac{11}{36}$.
15 Add 5, $7\frac{1}{2}$, $\frac{2}{5}$ and $\frac{7}{11}$ together. Res. $13\frac{59}{110}$.

NOTE 3.—When compound fractions are given, reduce them to single fractions, and proceed as before.

16 Add $\frac{9}{10}$ of $\frac{11}{12}$, $\frac{8}{11}$ of $\frac{6}{14}$, and $\frac{7}{15}$ of $\frac{5}{8}$ together. Res. $1\frac{643}{1540}$.

9240 common denominator

$\frac{9}{10}$ of $\frac{11}{12}=\frac{99}{120}$	$77\times99=7623$
$\frac{8}{11}$ of $\frac{6}{14}=\frac{48}{154}$	$60\times48=2880$
$\frac{7}{15}$ of $\frac{5}{8}=\frac{35}{120}$	$77\times35=2695$

$$\frac{13198}{9240}=1\frac{1970}{4620}.$$

17 Add $\frac{2}{4}$ of $\frac{7}{8}$ and $\frac{4}{6}$ of $\frac{10}{20}$ together. Res. $\frac{37}{48}$.
18 Add $1\frac{3}{5}$, $\frac{4}{5}$ of $\frac{1}{3}$, and $9\frac{3}{20}$ together. Res. $11\frac{1}{60}$.
19 Add $1\frac{9}{10}$, $6\frac{7}{8}$, $\frac{2}{3}$ of $\frac{1}{2}$, and $7\frac{1}{2}$ together. Res. $16\frac{73}{120}$.

NOTE 4.—When the given fractions are of several denominations, reduce them to their proper values or quantities, and add as in the following example.

20 Add $\frac{7}{9}$ of a pound, to $\frac{3}{10}$ of a shilling.
Result 15s. $10\frac{4}{15}$d.

	s.	*d.*	15
$\frac{7}{9}$ of a £	15	$6\frac{2}{3}$	5×2—10
$\frac{3}{10}$ of a *s.*	0	$3\frac{3}{5}$	3×3= 9
	15	$10\frac{4}{15}$	$\frac{19}{15}=1\frac{4}{15}$

21 Add $\frac{7}{8}$ of a pound, to $\frac{3}{4}$ of a shilling.
Result 18s. 3d.

22 Add $\frac{3}{4}$ of a penny, to $\frac{1}{9}$ of a pound.
Res. 2s. 3d. 1qr. $\frac{2}{3}$.

23 Add $\frac{1}{2}$ lb. troy, to $\frac{7}{12}$ of an ounce.
Res. 6oz. 11dwt. 16grs.

24 Add $\frac{3}{4}$ of a mile, to $\frac{7}{10}$ of a furlong. Res. 6fu. 28p.

25 Add $\frac{1}{2}$ of a yard, to $\frac{2}{3}$ of a foot. Res. 2ft. 2 in.

26 Add $\frac{1}{3}$ of a day, to $\frac{1}{2}$ of an hour. Res. 8h. 30min.

27 Add $\frac{1}{3}$ of a week, $\frac{1}{4}$ of a day, and $\frac{1}{2}$ of an hour together. Res. 2 days, 14 hours, 30min.

SUBTRACTION OF VULGAR FRACTIONS.

RULE.

Prepare the given fractions as in Addition; then subtract the less from the greater, and place the difference over the common denominator.

EXAMPLES.

1 Take $\frac{3}{5}$ from $\frac{4}{5}$. Rem. $\frac{1}{5}$.

2 Take $\frac{1}{12}$ from $\frac{7}{12}$. Rem. $\frac{1}{2}$.

3 Take $\frac{5}{10}$ from $\frac{9}{10}$. Rem. $\frac{2}{5}$.

4 Take $\frac{2}{5}$ from $\frac{3}{7}$. Rem. $\frac{1}{35}$.

	35
$\frac{3}{7}$	5×3=15
$\frac{2}{5}$	7×2=14
	$\frac{1}{35}$

$\frac{5}{7}$	$\frac{11}{12}$	$\frac{15}{16}$	$\frac{17}{35}$
$\frac{2}{7}$	$\frac{5}{12}$	$\frac{3}{16}$	$\frac{13}{35}$
$\frac{3}{7}$	$\frac{6}{12}=\frac{1}{2}$	$\frac{12}{16}=\frac{3}{4}$	$\frac{4}{35}$

12 com. denom

$\frac{11}{12}$ | $1\times11=11$
$\frac{2}{3}$ | $4\times 2=8$
$\frac{1}{4}$ Ans. $\frac{3}{12}=\frac{1}{4}$

60 com. denom.

$\frac{7}{15}$ | $4\times6=28$
$\frac{9}{20}$ | $3\times9=27$
$\frac{1}{60}$ Ans. $\frac{1}{60}$

$8\frac{2}{9}$	$9\frac{1}{4}$	$7\frac{7}{8}$	6	100
$6\frac{3}{11}$	$4\frac{3}{4}$	$3\frac{2}{3}$	$\frac{1}{8}$	$99\frac{99}{100}$
$1\frac{94}{99}$	$4\frac{9}{20}$	$4\frac{5}{24}$	$5\frac{7}{8}$	$\frac{1}{100}$

From $\frac{7}{9}$ of a pound take $\frac{3}{10}$ of a shilling.

15 com denom.

	s.	d.	
$\frac{7}{9}$ of a pound =	15	$6\frac{2}{3}$	$5\times2=10$
$\frac{3}{10}$ of a shilling =		$3\frac{3}{5}$	$3\times3=9$
	s. 15	$3\frac{1}{15}$	Ans. $\frac{1}{15}$

Fron. $\frac{3}{4}$ of a £ take $\frac{3}{4}$ of a shilling.
Res. 14s. 3d.

From $\frac{3}{4}$ of a lb. troy, take $\frac{1}{6}$ of an ounce.
Res. 8oz. 16dwt. 16grs.

From $\frac{1}{6}$ of a yard take $\frac{2}{3}$ of an inch. Res. 5in. $\frac{1}{3}$.

From $\frac{5}{9}$ of a £ take $\frac{2}{3}$ of $\frac{3}{4}$ of a shilling.
Res. 10s. 7d. 1qr. $\frac{1}{3}$.

MULTIPLICATION OF VULGAR FRACTIONS.

RULE.

Prepare the given fractions, if necessary; then multiply the numerators together for a new numerator, and the denominators together for a new denominator.

EXAMPLES.

1 Multiply $\frac{4}{9}$ by $\frac{3}{16}$. Res. $\frac{1}{12}$.

$$\frac{4}{9}\times\frac{3}{16}=\frac{12}{144}=\frac{1}{12}$$

2 Multiply $\frac{1}{12}$ by $\frac{3}{8}$. Res. $\frac{1}{32}$.
3 Multiply $\frac{6}{7}$ by $\frac{18}{6}$. Res. $2\frac{24}{42}$, or $2\frac{4}{7}$.
4 Multiply $12\frac{3}{5}$ by $7\frac{2}{3}$. Res. $96\frac{3}{5}$.

$12\frac{3}{5}=\frac{63}{5}$ and $7\frac{2}{3}=\frac{23}{3}$; then $\frac{63}{5}\times\frac{23}{3}=\frac{1449}{15}=96\frac{3}{5}$.

5 Multiply $7\frac{1}{4}$ by $8\frac{1}{2}$. Res. $61\frac{5}{8}$.
6 Multiply $4\frac{1}{2}$ by $\frac{1}{8}$. Res. $\frac{9}{16}$.
7 Multiply $\frac{7}{8}$ by $13\frac{9}{10}$. Res. $12\frac{13}{80}$.
8 Multiply $\frac{1}{3}$ of $\frac{4}{5}$ by $\frac{7}{10}$ of $\frac{11}{12}$. Res. $\frac{77}{450}$.
9 Multiply $4\frac{3}{4}$ by $\frac{2}{3}$ of $\frac{3}{4}$. Res. $2\frac{3}{8}$.
10 Multiply $\frac{1}{2}$ of 7 by $\frac{3}{6}$. Res. $1\frac{3}{4}$.
11 Multiply $2\frac{1}{3}$ by $1\frac{1}{7}$, and multiply the product by $\frac{1}{2}$ of $\frac{3}{4}$ of $\frac{2}{5}$. Res. $\frac{2}{3}$.

DIVISION OF VULGAR FRACTIONS.

RULE.

Prepare the given fractions, if necessary, then invert the divisor, and proceed as in Multiplication.

EXAMPLES.

1 Divide $\frac{4}{9}$ by $\frac{7}{8}$. Res. $\frac{32}{63}$.

$$\frac{8\times4=32}{7\times9=63}$$

2 Divide $\frac{4}{7}$ by $\frac{2}{3}$. Res. $\frac{6}{7}$.
3 Divide $\frac{17}{21}$ by $\frac{3}{5}$. Res. $1\frac{22}{63}$.
4 Divide $1\frac{1}{2}$ by $4\frac{8}{10}$. Res. $\frac{5}{16}$.
5 Divide $3\frac{1}{6}$ by $9\frac{1}{2}$. Res. $\frac{1}{3}$.
6 Divide $\frac{7}{8}$ by 4. Res. $\frac{7}{32}$.
7 Divide 4 by $\frac{7}{8}$. Res. $4\frac{4}{7}$.
8 Divide $\frac{1}{2}$ of $\frac{2}{3}$ by $\frac{2}{3}$ of $\frac{3}{4}$. Res. $\frac{2}{3}$.
9 Divide $\frac{1}{5}$ of 19 by $\frac{2}{3}$ of $\frac{3}{4}$. Res. $7\frac{3}{5}$.
10 Divide $4\frac{5}{9}$ by $\frac{5}{9}$ of 4. Res. $2\frac{1}{20}$.
11 Divide $\frac{8}{3}$ of $\frac{1}{3}$ by $\frac{5}{7}$ of $7\frac{3}{5}$. Res. $\frac{7}{171}$.
12 Divide $5205\frac{1}{5}$ by $\frac{4}{5}$ of 91. Res. $71\frac{1}{2}$.

PROPORTION IN VULGAR FRACTIONS.

RULE.

State the question, (as in page 89) reduce each term to its simplest form, invert the first term or terms, and proceed as in Multiplication of Vulgar Fractions.

EXERCISES.

1 If $\frac{3}{4}$ yd. cost $\$\frac{5}{8}$; what will $\frac{3}{5}$ yd. cost? Ans. 50c

yd. *yd.* *D.*

$\frac{3}{4} : \frac{3}{5} :: \frac{5}{8}$. Then $\frac{4}{3}\times\frac{3}{5}\times\frac{5}{8}=\frac{60}{120}=\$\frac{1}{2}=50$ *cts.*

2 If $\frac{1}{4}$ bu. cost $\$\frac{7}{8}$; what will $\frac{4}{5}$ bu. cost? Ans. \$2,80.

bu. *bu.* *D.*

$\frac{1}{4} : \frac{4}{5} :: \frac{7}{8}$. Then $\frac{4}{1}\times\frac{4}{5}\times\frac{7}{8}=\frac{112}{40}=\$2\frac{32}{40}=\$2,80$.

3 If A owned $\frac{3}{5}$ of a toll-bridge, and sold $\frac{3}{4}$ of his share for \$684; what is the whole value? Ans. \$1520.

$\frac{3}{4}$ of $\frac{3}{5} : \frac{1}{1} :: \frac{684}{1}$: that is, $\frac{9}{20} : \frac{1}{1} :: \frac{684}{1}$. Then $\frac{20}{9}\times\frac{1}{1}\times\frac{684}{1}=\frac{13680}{9}=\$1520$.

4 If I barter $5\frac{8}{9}$ cwt. of sugar at $6\frac{3}{4}$ cts. per lb., for indigo at $\$4\frac{5}{16}$ per lb.; how much indigo must I receive? Ans. 10 lb. 5 oz. $2\frac{4}{5}$ dr.+

D. *cts.* *cwt.* *cts.* *cts.* *lb.*

$4\frac{5}{16} : 6\frac{3}{4} :: 5\frac{8}{9}$; that is $\frac{6900}{16} : \frac{27}{4} :: \frac{5936}{9}$. Then

16 4 9 $\frac{16}{6900}\times\frac{27}{4}\times\frac{5936}{9}=\frac{2564352}{248400}=10$ lb. 5 oz. 2 dr. $\frac{4191}{5175}$.

$\frac{69}{16}$ $\frac{27}{4}$ $\frac{53}{9}$

5 If the cent roll weighs $6\frac{1}{4}$ oz., when wheat is $68\frac{3}{4}$ cents per bu.; what is the cost of wheat per bu. when it weighs $4\frac{1}{6}$ oz? Ans. $\$1,03\frac{1}{8}$.

oz. *oz.* *cts.*

$4\frac{1}{6} : 6\frac{1}{4} :: 68\frac{3}{4}$; then $\frac{6}{25}\times\frac{25}{4}\times\frac{275}{4}=\frac{41250}{400}=103\frac{1}{8}$.

6 4 4

$\frac{25}{6}$ $\frac{25}{4}$ $\frac{275}{4}$

6 How many men will reap $417\frac{3}{5}$ acres in $12\frac{1}{2}$ days, if 5 men reap $52\frac{1}{5}$ in $6\frac{1}{4}$ days? Ans. 20.

a. *a.* *men.*

$52\frac{1}{5} : 417\frac{3}{5}$ / *da.* *da.* / $12\frac{1}{2} : 6\frac{1}{4}$ $:: 5$; that is $\frac{261}{5} : \frac{2088}{5}$ / $\frac{25}{2} : \frac{25}{4}$ $:: \frac{5}{1}$. Then

$\frac{5}{261}\times\frac{2}{25}\times\frac{2088}{5}\times\frac{25}{4}\times\frac{5}{1}=\frac{20880}{1044}=20$.

NOTE.—In multiplying, omit the numbers that are in both the upper and lower series.

DECIMAL FRACTIONS.

A decimal fraction is a fraction whose denominator is 1, with as many cyphers annexed as there are figures in the numerator, and is usually expressed by writing the numerator only with a point prefixed to it: thus $\frac{5}{10}$, $\frac{75}{100}$, $\frac{625}{1000}$, are decimal fractions, and are expressed by .5, .75, .625.

A mixed number, consisting of a whole number and a decimal, as $25\frac{5}{10}$, is written thus, 25.5.

As in numeration of whole numbers the values of the figures increase in a tenfold proportion, from the right hand to the left; so in decimals, their values decrease in the same proportion, from the left hand to the right, which is exemplified in the following

TABLE.

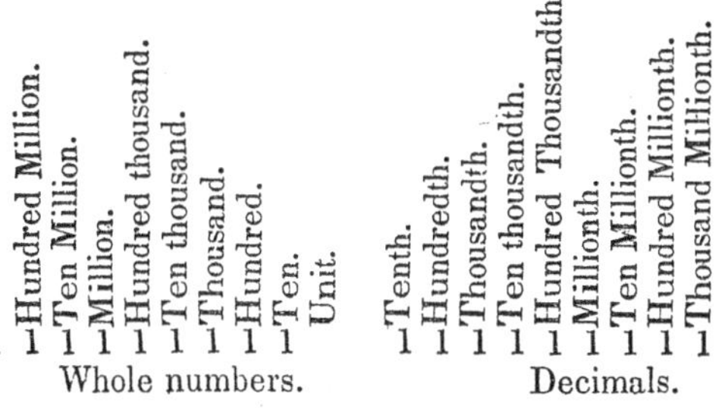

NOTE.—Cyphers annexed to Decimals, neither increase nor decrease their value; thus, .5, .50, .500, being $\frac{5}{10}$, $\frac{50}{100}$, $\frac{500}{1000}$, are of the same value: but cyphers prefixed to decimals, decrease them in a tenfold proportion; thus .5, .05, .005; being $\frac{5}{10}$, $\frac{5}{100}$, $\frac{5}{1000}$, are of different values.

ADDITION OF DECIMALS.

RULE.

Place the given numbers according to their values, viz. units under units, tenths under tenths, &c., and add as in addition of whole numbers; observing to set the point in the sum exactly under those of the given numbers.

EXAMPLES.

.12	2.16	.14	.1	.15
.134	3.45	.24	4.12	.75
.21	40.02	.122	15.4	.92
743	35.4	.36	76.36	63.25
345	36.1	.141	120.16	25.
.002	125.32	.567	425.04	4.
1.554	242.45			

6 Add .5, .75, .125, 496, and .750 together.

7 Add .15, 126.5, 650.17, 940.113, and 722.2560 together.

8 Add 420., 372.45, .270, 965.02, and 1.1756 together.

SUBTRACTION OF DECIMALS.

RULE.

Place the numbers as in addition, with the less under the greater, and subtract as in whole numbers; setting the point in the remainder under those in the given numbers.

EXAMPLES.

.4562	56.12	.4314	5672.1	32.456
.316	1.242	.312	321.12	1.33
.1402	54.878			

6 From 100.17 take 1.146.

7 From 146.265 take 45.3278.

8 From 4560. take .720.

MULTIPLICATION OF DECIMALS.

RULE.

Multiply as in whole numbers, and from the right hand of the product, separate as many figures for decimals, as there are decimal figures in both the factors.

EXAMPLES.

1 Multiply .612 by 4.12

```
   .612
   4.12
   ----
   1224
   612
 2448
 ------
 2.52144
```

2 Multiply 1.007 by .041.

```
   1.007
    .041
   -----
    1007
   4028
  -------
  .041287
```

3	Multiply	37.9 by 46.5	Product 1762.35
4		36.5 by 7.27	265.355
5		29.831 by .952	28.399112
6		3.92 by 196.	768.32
7		.285 by .003	.000855
8		4.001 by .004	.016004
9		.00071 by .121	.00008591

DIVISION OF DECIMALS.

RULE.

Divide as in whole numbers, and from the right hand of the quotient, separate as many figures for decimals as the decimal figures of the dividend exceed those of the divisor. If there are not so many figures as the rule requires, supply the defect by prefixing cyphers.

EXAMPLES.

1 Divide .863972 by .92

```
.92).863972(.9391
     828
     ---
      359
      276
      ---
       837
       828
       ---
         92
         92
         --
```

2 Divide 4.13 by 572.4.

```
572.4)4.130000(.00721+
       40068
       -----
        12320
        11448
        -----
          8720
          5724
          ----
          2996
```

3	Divide 19.25 by 38.5	Quotient	.5
4	234.70525 by 64.25		3.653
5	1.0012 by .075		13.34+
6	.1606 by .44		.365
7	.1606 by 4.4		.0365
8	.1606 by 44.		.00365
9	9. by .9		10.
10	.9 by 9.		.1
11	186.9 by 7.476		25.

NOTE 1. When a whole number is to be divided by a greater whole number, cyphers must be affixed to the dividend, as decimal figures.

12	Divide 3 by 4	Quotient	.75
13	275 by 3842		.071577+
14	210 by 240		.875

NOTE 2. When any whole number is divided by another, if there be a remainder, cyphers may be affixed to the dividend, and the quotient continued.

15	Divide 382 by 25	Quotient	15.28
16	13689 by 75		182.52
17	315 by 124		2.5403+

REDUCTION OF DECIMALS.

CASE 1.

To reduce a vulgar fraction to a decimal.

RULE.

Place cyphers to the right of the numerator, until you can divide it by the denominator, and continue to divide until there is no remainder left; or if it be a number which will never come out without a remainder, until it is carried out to a convenient number of decimal places.

EXAMPLES.

1 Reduce $\frac{4}{5}$ to a decimal.

```
5)40
 ---
 .8 Ans.
```

2 Reduce $\frac{7}{8}$ to a decimal. Ans. .875.
3 Reduce $\frac{17}{24}$ to a decimal. Ans. .70833.+
4 Reduce $\frac{381}{2162}$ to a decimal. Ans. .1762.+
5 Reduce $\frac{116}{254}$ to a decimal. Ans. .4566.+

CASE 2.

To reduce any given sum or quantity to the decimal of any higher given denomination.

RULE.

Reduce the given sum or quantity to the lowest denomination mentioned in it.

Reduce one of that denomination of which you wish to make it a decimal, to the same denomination with the given sum.

Divide the given quantity so reduced by one of the denomination of which you wish to make it a decimal; the quotient will be the decimal required.

EXAMPLES.

1 Reduce 3s. 6d. to the decimal of a pound.

```
3s. 6d.= 42      240)42.000(.175 decimals. Ans.
£1.    =240          240
                     ----
                     1800
                     1680
                     ----
                      1200
                      1200
                      ----
```

2 Reduce 2R. 4P. to the decimal of an acre. Answer, .525.

3 Reduce 2 qr. 2 nails to the decimal of a yard. Ans. .625.

4 Reduce 5 minutes to the decimal of an hour. Ans. .08333.

5 Reduce 10 grains to the decimal of an ounce, apothecaries' weight. Ans. .02083.+

6 Reduce 2 quarts 1 pint to the decimal of a hogshead. Ans. .00992.+

CASE 3.

To reduce a decimal fraction to its proper value.

RULE.

Multiply the given fraction continually by the denomination next lower than that of which it is a decimal, for the proper value.

EXAMPLES.

1 What is the value of .375 of a dollar? Ans. 37½cts.

```
 .375
  100
 ------
37.500
    10
 ------
 5.000
```

2 What is the value of .1361 of a £.? Ans. 2s. 8½d.

3 What is the value of .235 of a day?
Ans. 5 hours, 38 min. 24 sec.

4 What is the value of .42 of a gallon?
Ans. 1 quart, 1.36 pt.

5 What is the value of .253 of a shilling? Ans. 3.036d.

6 What is the value of .436 of a yard?
Ans. 1 qr. 2.976 na.

7 What is the value of .9 of an acre?
Ans. 3R. 24P.

PROPORTION IN DECIMALS.

RULE.

State the question as the rule of three, in whole numbers, only observe, when you multiply and divide, to place the decimal points according to the rules of multiplication and division of decimals.

EXAMPLES.

1 If 4.2lb of coffee cost 8s. 2.3d., what cost 639.25lb.?

lb.	*lb.*	*s. d.*	*£ s. d.*	
4.2 :	639.25 : :	8 2.3 :	62 6 9.49	Ans.

2 When 1.4 yard cost 13s. what will 15 yards come to at the same price? Ans. £6 19s. 3d. 1.71 qr.

3 If I sell 1 qr. of cloth for 2 dollars 34.5 cents, what is it per yard? Ans. $9 38 cts.

4 A merchant sold 10.5 cwt. of sugar, for 108.30 dollars, for which he paid 84 dollars 39.12 cents; what did he gain per cwt. by the sale? Ans. $2 27 cts. 7m.+

5 How many pieces of cloth, at 20.8 dollars per piece, are equal in value to 240 pieces, at 12.6 dollars per piece? Ans. 145.38+ pieces.

6 If, when the price of wheat is 74.6 cents per bushel, the penny roll weighs 5.2 oz., what should it be per bushel when the penny roll weighs 3.5 oz.?
Ans. $1 10 cts. 8m.+

Question.

How do you perform operations in the rule of three in decimals?

COMPOUND PROPORTION, IN DECIMALS.

Questions in this rule are wrought as in whole numbers, placing the points agreeably to former directions.

EXAMPLES.

1 If 3 men receive 8.9£ for 19.5 days labor, how much must 20 men have for 100.25 days?

Ans. 305£. 0s. 8.2d.

men	3 : 20	. : 89£. : 305£. 0s. 8.2d.
days	19.5 : 100.25 days	

2 If 2 persons receive 4.625s. for 1 day's labor, how much should 4 persons have for 10.5 days?

Ans. 4£. 17s. 1½d.

3 If the interest of 76.5£ for 9.5 months, be 15.24£. what sum will gain 6£ in 12.75 months?

Ans. 22£ 8s. 9¾d.

4 How many men will reap 417.6 acres in 12 days, if 5 men reap 52.2 acres in 6 days? Ans. 20 men.

5 If a cellar 22.5 feet long, 17.3 feet wide, and 10.25 feet deep, be dug in 2.5 days, by 6 men, working 12.3 hours a day, how many days of 8.2 hours, should 9 men take to dig another, measuring 45 feet long, 34.6 wide, and 12.3 deep? Ans. 12 days.

MENSURATION.

Mensuration is employed in measuring masons' and carpenters' work, plastering, painting and paving; also, for measuring timber in all its forms, and for estimating quantity in length, superfices, and solids, whenever yards, feet, inches, &c., are employed.

The denominations are, foot, inch, second, third, and fourth,

12 Fourths	one 1 Third‴
12 Thirds	one 1 Second″
12 Seconds	one 1 Inch. *I.*
12 Inches	one 1 Foot. *Ft.*

ADDITION.

RULE.

Proceed as in Compound Addition.

EXAMPLES.

Ft.	I.	″		Ft.	I.	″		Ft.	I.	″	‴	⁗
25	0	3		72	4	0		17	9	2	3	11
14	2	9		54	3	2		18	11	10	8	9
35	11	10		14	0	8		22	11	5	4	9
45	10	11		26	3	2		14	10	11	10	8
6	0	0		19	0	4		12	0	0	4	10
4	9	0		14	0	0		10	2	8	4	0
132	4	9										

4 Four floors in a certain building contain each 1084 feet, 9in. 8″; how many feet are there in all?

Ans. 4339ft. 2 in. 8″.

5 There are six mahogany boards, the first measures 27 ft. 3in., the second 25 ft. 11in., the third, 23ft. 10in., the fourth 20ft. 9in., the fifth 20ft. 6in., and the sixth 18 feet 5 in.; how many feet do they contain?

Ans. 136ft. 8in.

SUBTRACTION.

RULE.

Proceed as in Compound Subtraction.

EXAMPLES.

Ft.	I.	″		Ft.	I.	″		Ft.	I.	″	‴	⁗
75	9	9		84	6	4		100	10	8	10	11
14	6	11		72	9	8		97	2	4	6	8
61	2	10										

4 If 19ft. 10in. be cut from a board which contains 41ft. 7in. how much will be left? Ans. 21ft. 9in.

5 Bought a raft of boards containing 59621ft. 8in., of which are since sold 3 parcels, each 14905ft. 5in.; how many feet remain? Ans. 14905ft. 5in.

MULTIPLICATION.

Case 1.

When the feet of the multiplier do not exceed 12.

RULE.

Set the feet of the multiplier under the lowest denomination of the multiplicand, as in the following example; then multiply as in Compound Multiplication, by each denomination of the multiplier separately, observing to place the right hand figure, or number, of each product, under that denomination of the multiplier by which it is produced.

EXAMPLES.

1 Multiply 10 feet 6 inches by 4 feet 6 inches.
Product 47 feet, 3 in

Ft.	*I.*	''
10	6	
	4	6
5	3	0
42	0	
47	3	0

A table 10 feet 6 inches long, and 1 foot wide, will make 10 feet 6 inches, or 10½ feet, square measure.

And 4 feet 6 inches, or 4½ feet wide, will make 4½ times 10½, or 47¼ feet, or 47 feet 3 inches.

OR THUS:

ft.	*in.*
10	6
	4½ ft.
5	3
42	0
47	3

Note 1.—If there are no feet in the multiplier, supply their place with a cypher; and in like manner supply the place of any other denomination between the highest and lowest.

10ft. 6in. or 10½ feet long.

4½ wice

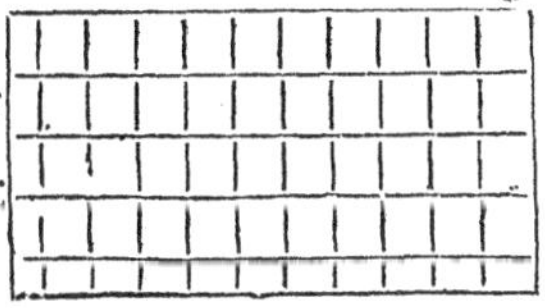

4 times 10½ make 42ft.
and ½ time 10½ make 5¼ ft.
Added, make 47¼.

		Ft.	I.	"		Ft.	I.			Ft.	I.	"	'''
2	Multiply	9	7		by	3	6		Res.	33	6	6	
3		3	11		by	9	5			36	10	7	
4		8	6	9	by	7	3	8		62	6	7	9
5		28	10	6	by	3	2	4		92	2	10	6

CASE 2.

When the feet of the multiplier exceed 12.

RULE.

Multiply by the feet of the multiplier as in Compound Multiplication, and take parts for the inches, &c.

EXAMPLES.

1 Multiply 112ft. 3in. 5″ by 42ft. 4in. 6″

```
             Ft.   I.  "
             112   3   5
                       6×7=42
             ------------
             673   8   6
                       7
   I.        ------------
 | 4 | ⅓ |  4715  11   6  '''  ''''
 | " |   |    37   5   1   8
 | 6 | ⅛ |     4   8   1   8   6
             --------------------
            4758   0   9   4   6
```

		Ft.	I.	"		Ft.	I.	"		Ft.	I.	"	'''
2	Multiply	76	7		by	19	10		Res.	1518	10	10	
3		127	6		by	184	8			23545	0	0	
4		71	2	6	by	81	1	8		5777	9	2	2

APPLICATION.

1 A certain board is 28ft. 10in. 6'' long, and 3ft. 2 in. 4'' wide; how many square feet does it contain?

Ans. 92ft. 2in. 10'' 6'''.

2 If a board be 23ft. 3in. long, and 3ft. 6in. wide, how many square feet does it contain?

Ans. 81ft. 4in. 6''.

3 A certain partition is 82ft. 6in. by 13ft. 3in.; how many square feet does it contain? Ans. 1093ft. 1in. 6''.

4 If a floor be 79ft. 8in. by 38ft. 11in., how many square feet are therein? Ans. 3100ft. 4in. 4''.

NOTE.—Divide the square feet by 9, and the quotient will be square yards.

5 If a ceiling be 59ft. 9in. long, and 24ft. 6in. broad, how many square yards does it contain?

Ans. 162yd. 5ft.+

		Ft.	I.	
6 in.	½	59	9	
			3	
		179	3	
			8	
		1434	0	''
		29	10	6
		9)1463	10	6
		162yd.	5 ft.	

6 How many yards are contained in a pavement 56 feet 9 inches long, and 18 feet 4 inches wide?

Ans. 115yd. 5ft 5in.

7 How many yards in a ceiling 92ft. 4in. long, 22ft. 8in. wide? Ans. 232yd. 4ft. 10in.+

8 How many squares in a floor 37ft. 6in. long, and 21ft. 9in. wide? Ans. 8 squares, 15 feet.+

A square is 10 feet long and 10 feet wide, or 100 square feet. It is used in estimating flooring, roofing, weather-boarding, &c.

9 How many squares of weather-boarding on the side of a house 43 feet 6 in. long, and 18ft. 8in. high?

Ans. 8 squares, 12ft.

10 How many squares in a roof 36ft. 4in. long, 15ft. 9in. wide? Ans. 5 sq. 72ft.+

Note 2.—*To measure a triangle.* Multiply the base by one half the perpendicular height, and the product will be its superficial content.

11 Let C, H, and G, represent a triangle, whose base is 40 feet, and perpendicular height 28 feet; how many feet does it contain? Ans. 560 feet.

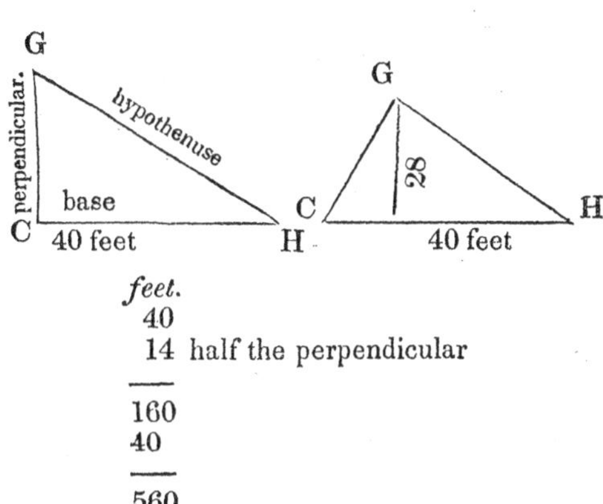

feet.
40
14 half the perpendicular
———
160
40
———
560

12 How many square feet in a triangle 80 feet long and 36 feet high? Ans. 1440 ft.

13 In a triangular pavement 46 feet long, and 24 feet at the place of its greatest width, how many yards; and how many bricks, allowing 41 to every square yard?
Ans. 61yd. 3ft., and 2514 bricks.

14 In the gable ends of a house, which is 63 feet long and 22 feet high, from the "square of the building" to the top, how many squares? Ans. 6sq. 93in.

Note 3.—*To find the circumference of a circle, when the diameter is given:* Say,

As 7 are to 22, so is the diameter to the circumference; *or the contrary,*

As 22 are to 7, so is the circumference to the diameter.

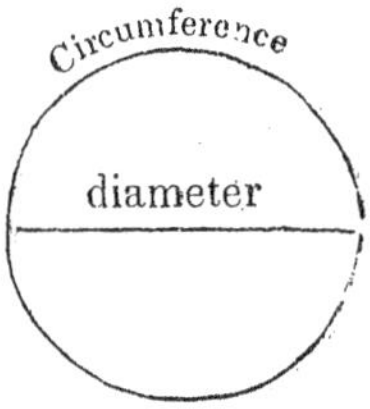

The diameter of a circle is 14 feet; what is the circumference? Ans. 44.

As 7 : 22 : : 14 : 44

The circumference of a circle is 44ft.; what is the diameter? Ans. 14.

Circumference
diameter

NOTE 4.—*To find the superficial contents of a circle* Multiply half the circumference by half the diameter

15 How many square feet in a circle whose diameter is 14 feet, and circumference 44? Ans 154 ft.

half circumference, 22
half diameter, 7

154 feet

16 How many square feet in a circle whose circumference is 16 feet? Ans. 20 sq. ft.

As 22 : 7 : : 16 : 5

half cir. 8
half diam. 2½
—20

17 How many square feet in a circle whose diameter is 21 feet? Ans. 346½ ft.

NOTE 5.—*To find the superficial contents of a globe.* Multiply the circumference by the diameter.

18 What are the superficial contents of a globe whose diameter is 70 feet, and circumference 220 feet? Ans. 15400 sq. ft.

19 How many square feet of cloth would be required to cover a globe, whose diameter is 28 feet, and circumference 88? Ans. 2464 ft.

20 How many yards of canvass would be required to make a balloon of a globular form, 20yardsin diameter? Ans. 1257 sq. yds.

NOTE 6.—*To find the solid contents of a cube,* or of a square stick of timber, or a pile of wood, &c.* Multiply the length by the breadth, and that product by the thickness.

* A *cube* is a solid body, contained by 6 equal sides, all of which are exact squares.

21 What are the solid contents of a cube whose diameter is 4 feet? Ans. 64 feet.

4 feet
4
—
16
4
—
64

22 What is the solid contents of a stick of timber 2 feet thick, 3 feet wide, and 36 feet long?
Ans. 216 solid feet.

36 feet
3
—
108
2
—
216

23 How many solid feet in a block of marble 3 feet thick, 7 feet wide, and 13 feet long? Ans. 273 sol. ft.

24 In a cube whose diameter is 7 feet, how many solid feet? Ans. 343 feet.

25 How many solid feet in a pile of wood 28 feet long, 8 feet wide, and 10 feet high; and how many cords does it contain? Ans. 2240 feet; 17 cords 64 ft. or, $17\frac{1}{2}$ cords.

26 In a cellar 36 feet long, 27 feet wide, and $4\frac{1}{2}$ feet deep, how many solid yards? Ans. 162 yards.

27 How many perches* of stone in a wall 42 feet long, $8\frac{1}{2}$ feet high, and 2 feet thick? Ans. 28,8 per.

feet
24,75)714.00(28,8

28 In a 12 inch brick wall, 52 feet long and 36 feet high, how many bricks, allowing 21 to every square foot of wall? Ans. [illegible]9312.

---

* A perch is $16\frac{1}{2}$ feet long, $1\frac{1}{2}$ ft. wide, and 1 foot hig[illegible] $24\frac{3}{4}$ solid feet.

29 In an 8 inch brick wall, 82 feet long and 16 feet high; how many bricks, allowing 14 bricks for every square foot of wall? Ans. 18368.

30 In a 16 inch brick wall, 148 feet long and 42 feet high, how many bricks, allowing 28 bricks to the square foot? Ans. 174048.

31 How many bricks in 3 walls, the first 68 feet long, 18 feet 6 inches high, 16 inches thick; the second 72 ft. 6in. long, 19ft. 4in. high, 12in. thick; the third 43ft. 4in. long, 12ft. 8in. high, 8in. thick? Ans. 72343.+

NOTE 7.—*To find the solid contents of a cylinder.**— Find the contents of one end by *Note* 4, and multiply that product by the length.

32 What are the solid contents of a cylinder whose diameter is 14 feet, and length 16 feet? Ans. 2464ft.

half circumference	22
half diameter	7
	154
	16
	924
	154
	2464 feet

33 What are the contents of a circular well, 7 feet in diameter, and 62 feet deep? Ans. 2387 ft.

34 What are the solid contents of a tub whose diameter is 6 feet and height 7 feet? Ans. 198 ft.

35 How many feet in a circular well, 10 feet diameter and 20 feet deep? Ans. $157\frac{3}{7}$ feet.

NOTE 8.—*To find the solid contents of the frustrum of a cone,* To the sum of the squares of the two diameters, in inches, add their product; multiply this sum by one-third of the depth, and this last product by the decimal .7854. Eu. El.

The result will be the contents in cubic inches, which reduce as desired.

* A Cylinder is a long round body, whose diameter is every where the same.

The liquid gallon of Ohio contains 231 cubic inches.
The dry gallon contains $268\frac{4}{5}$ cubic inches.
The bushel, of grain, contains $2150\frac{2}{5}$ cubic inches.
The bushel, of coal, lime, &c. 2688 cubic inches.

EXAMPLES.

1 In a circular vessel, whose greater diameter is 80 inches, the less 71, and the depth 34, what is the contents in liquid gallons; and also in bushels of grain?

Ans. { 659.72+ gallons.
70.86 bushels.

80×80=6400= square of 80.
71×71=5041= square of 71.

11441
80×71=5680= product of 80 and 71.

17121
$\frac{1}{3}$ in depth = $11\frac{1}{3}$=$\frac{1}{3}$ of the depth.

194038×.7854=152397.4452 inches.

Divide by 231 for gallons, and $2150\frac{2}{5}$ for bushels.

2 The greater diameter of a tub is 38 inches, the less 20.2, and the depth 21; what is the content in gallons? Ans. 62.34+gallons.

3 The top diameter of a tube is 22 inches, the bottom 40, and the height 60; what is its contents in gallons, also in bushels of grain? Ans. { 201.55+gals.
21.64+bu.

4 How many barrels, of 32 gallons each, in a cistern, whose greater diameter is 8 feet 6 inches, the less 8 feet, and depth 7 feet 9 inches? Ans. 96 barrels 28+gals.

5 How many bushels of grain in a bin that is 8 feet long, 4 feet wide, and 6 feet high? Ans. 123+bu.

6 How many bushels of coal in a boat 60 feet long, 16 feet wide, and $4\frac{1}{2}$ feet deep? Ans 2777+bushels.

Note 9.—*To find the solid ccntents of a round stick of timber of a taper from one end to the other.* Find the circumference a little nearer the larger than the smaller end; from this, by *Note* 3, find the diameter: multiply half the diameter by half the circumference, and the product by the length.*

EXAMPLES.

1 What are the solid contents of a round stick of timber 10 feet long, and 2.61 feet circumference?

Ans, 5.4 feet.+

As 22 : 7 : : 2.61 : .83 diameter
1,305 half circumference
.415 half diameter

6525
1305
5220

.541575
10 length

5.415750

2 How many solid feet in a log 40 feet long, which girts 66 inches? Ans. 96.25 ft.

As 22 : 7 : : 66 : 21 in diameter

$33\times10\frac{1}{2}\times40=13860.$ 144)13860(96.25

Note 10.—*To find the solid contents of a globe.*—Multiply the cube of the diameter by .5236.

EXAMPLES.

1 What are the solid contents of a globe whose diameter is 14 inches? Ans. 1436.75in.+

$14\times14\times14=2744.$ $2744\times.5236=1436.7584.$

2 What are the contents of a balloon of a globular form, 42 feet in diameter? Ans. 38792.4 ft.+

3 How many solid miles are contained in the earth, or globe, which we inhabit?

* This method, though not quite accurate, is sufficiently near the truth for the purpose of measuring timber.

Suppose the diameter to be 7954 miles; then, 7954× 7954×7954=503218686664 the cube of the earth's axis, or diameter; then,

503218686664×.5236=263485304337 cubick miles. Ans.

NOTE.—The solidity of a globe may be found by the circumference, thus—Multiply the cube of the circumference by .016887—the product will be the contents.

INVOLUTION, OR THE RAISING OF POWERS.

The product arising from any number multiplied by itself, any number of times, is called its *power*, as follows:

2×2= 4 the square, or 2d power of 2.
2×2×2= 8 3d power or cube of 2.
2×2×2×2=16 4th power of 2.

The number which denotes a power is called its index.

NOTE.—When any power of a vulgar fraction is required, first raise the numerator to the required power, and then the denominator to the required power, and place the numerator over the denominator as before: thus, the 4th power of $\frac{2}{3}$ $\frac{2\times2\times2\times2}{3\times3\times3\times3}=\frac{16}{81}$

Questions.

What is the product, arising from the multiplication of any figure by itself a given number of times, called?

What is the number which denotes a power, called?

How do you proceed to find any required power of a vulgar fraction?

Table of the first nine Powers.

Roots.	*Squares.*	*Cubes*	*4th power.*	*5th power.*	*6th power.*	*7th power.*	*8th power.*	*9th power.*
1	1	1	1	1	1	1	1	1
2	4	8	16	32	64	128	256	512
3	9	27	81	243	729	2187	6561	19683
4	16	64	256	1024	4096	16384	65536	262144
5	25	125	625	3125	15625	78125	390625	1953125
6	36	216	1296	7776	46656	279936	1679616	10077696
7	49	343	2401	16807	117649	823543	5764801	40353607
8	64	512	4096	32768	262144	2097152	16777216	134217728
9	81	729	6561	59049	531441	4782969	43046721	387420489

EXAMPLES.

1 What is the square of 32?

```
 32
 32
 --
 64
96
----
1024 Ans
```

2 What is the cube of 14? Ans. 2744.

```
  14
  14
  --
  56
 14
 ---
 196
  14
 ---
 784
196
----
2744
```

3 What is the sixth power of 2.8? Ans 481,890304

4 What is the third power of .263? Ans. .018191447.

EVOLUTION, OR THE EXTRACTING OF ROOTS.

The root of a number, or power, is such a number, as being multiplied into itself a certain number of times, will produce that power, Thus 2 is the square root of 4, because $2\times2=4$; and 4 is the cube root of 64, because $4\times4\times4=64$, and so on.

THE SQUARE ROOT.

The square of a number is the product arising from that number multiplied into itself.

Extraction of the square root is the finding of such a number, as being multiplied by itself, will produce the number proposed. Or, it is finding the length of one side of a square.

RULE.

1 Separate the given number into periods of two figures, each, beginning at the units place.

2 Find the greatest square contained in the left hand period, and set its root on the right of the given number: subtract said square from the left hand period, and to the remainder bring down the next period for a dividual.

3 Double the root for a divisor, and try how often this divisor (with the figure used in the trial thereto annexed) is contained in the dividual: set the number of times in the root; then, multiply and subtract as in division, and bring down the next period to the remainder for a new dividual.

4 Double the ascertained root for a new divisor, and proceed as before, till all the periods are brought down.

NOTE.—If, when all the periods are brought down, there be a remainder, annex cyphers to the given number, for decimals, and proceed till the root is obtained with a sufficient degree of exactness.

Observe that the decimal periods are to be pointed off from the decimal point toward the right hand: and that there must be as many whole number figures in the root, as there are periods of whole numbers, and as many decimal figures as there are periods of decimals.

PROOF.

Square the root, adding in the remainder, (if any,) and the result will equal the given number.

EXAMPLES.

1 What is the square root of 5499025?

```
        5,49,90,25(2345 Ans.
        4                              2345
        ----                           2345
     43)149                            -----
        129                            11725
        ----                           9380
    464)2090                          7035
         1856                        4690
         ----                        -------
   4685)23425                        5499025 Proof.
        23425
```

2 What is the square root of 106929? Ans. 327.

3 What is the square root of 451584? Ans. 672.

4 What is the square root of 36372961? Ans. 6031.

5 What is the square root of 7596796? Ans. 2756.228+

6 What is the square root of 3271.4007? Ans. 57.19+

7 What is the square root of 4.372594? Ans. 2.091+

8 What is the square root of 10.4976? Ans. 3.24

9 What is the square root of .00032754? Ans. .01809+

10 What is the square root of 10? Ans. 3.1622+

To extract the Square Root of a Vulgar Fraction.

RULE.

Reduce the fraction to its lowest terms, then extract the square root of the numerator for a new numerator, and the square root of the denominator for a new denominator.

NOTE.—If the fraction be a surd, that is, one whose root can never be exactly found, reduce it to a decimal, and extract the root therefrom.

EXAMPLES.

1 What is the square root of $\frac{7056}{9216}$? Ans. $\frac{7}{8}$.

2 What is the square root of $\frac{2704}{4225}$? Ans. $\frac{4}{5}$.

3 What is the square root of $\frac{478}{549}$? Ans. .93309+

To extract the Square Root of a Mixed Number.

RULE.

Reduce the mixed number to an improper fraction, and proceed as in the foregoing examples: or,

Reduce the fractional part to a decimal, annex it to the whole number, and extract the square root therefrom.

EXAMPLES.

1 What is the square root of $37\frac{36}{49}$? Ans. $6\frac{1}{7}$.

2 What is the square root of $27\frac{9}{16}$? Ans. $5\frac{1}{4}$.

3 What is the square root of $85\frac{14}{15}$? Ans. 9.27+

APPLICATION.

1 The square of a certain number is 105625: what is that number? Ans. 325.

2 A certain square pavement contains 20736 square stones, all of the same size; what number is contained in one of its sides? Ans. 144.

3 If 484 trees be planted at an equal distance from each other, so as to form a square orchard, how many will be in a row each way? Ans. 22.

4 A certain number of men gave 30s. 1d. for a charitable purpose; each man gave as many pence as there were men: how many men were there? Ans. 19.

5 The wall of a certain fortress is 17 feet high, which is surrounded by a ditch 20 feet in breadth; how long must a ladder be to reach from the outside of the ditch to the top of the wall? Ans. 26.24+feet.

NOTE.—The square of the longest side of a right angled triangle is equal to the sum of the squares of the other two sides; and consequently, the difference of the square of the longest, and either of the other, is the square of the remaining one.

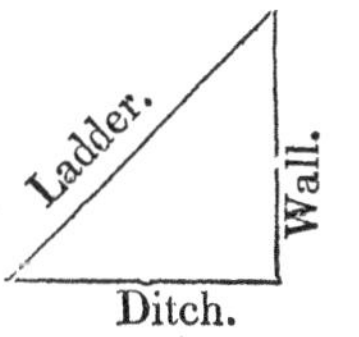

Ladder.
Wall.
Ditch.

6 A certain castle which is 45 yards high, is surrounded by a ditch 60 yards broad; what length must a ladder be to reach from the outside of the ditch to the top of the castle? Ans. 75 yards.

7 A line 27 yards long, will exactly reach from the top of a fort to the opposite bank of a river, which is known to be 23 yards broad; what is the height of the fort? Ans. 14.142+yards.

8 Suppose a ladder 40 feet long be so planted as to reach a window 33 feet from the ground, on one side of the street, and without moving it at the foot, will reach a window on the other side 21 feet high; what is the breadth of the street? Ans. 56.64+feet.

9 Two ships depart from the same port; one of them sails due west 50 leagues, the other due south 84 leagues; how far are they asunder??

Ans. 97.75+ Or, $97\frac{3}{4}$+leagues.

Questions.

What is a *square?* A square is a surface whose length and breadth are equal, and whose angles (or corners) are right angles, (or square.)

What is its *square root?* The square root is the length of the side of a square.

If the square be *sixteen,* what is the root?

Why is the root four?

If the root be *three,* what is the SQUARE?

What is the *square root* of twenty-five?

What is the square of five?

What is the square root of thirty-six?

What is the square of six?

How do you point off a number whose square root is to be extracted?

What is the next step? What do you subtract from the period? What do you annex to the remainder?

Illustration of the Rule for extracting the Square Root.

The reason for pointing off the given number into periods of *two* figures each, is, that the product of any whole number contains just as many figures as are in both the multiplier and the multiplicand, or but *one* less; consequently, the square contains just double as many figures as the root, or *one* less.

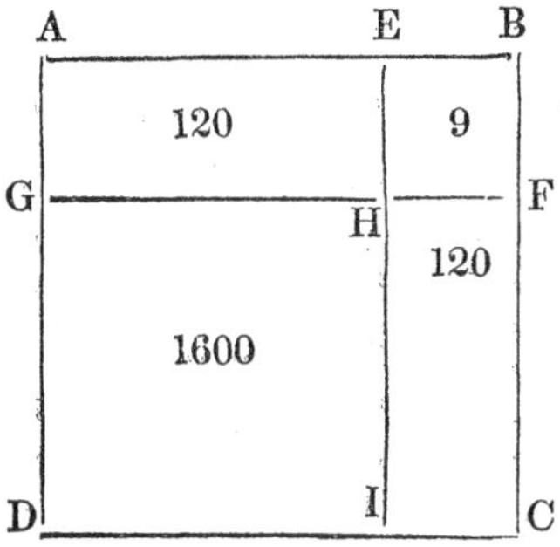

A E B

120	9
1600	120

G H F

D I C

Suppose the figure ABCD contains 1849 square feet, and that the number consists of two periods; then there must be two figures in the root.

The largest root whose square can be taken out of the left hand period, is 4, (or as it will stand in ten's place in the root, it is 40,) and the square of this is 16 (or 1600.) This taken from the whole square ABCD, or 1849, leaves 249.

```
   18,49(43
   16
   ---
83)249
   249
   ---
   ...
```

Now double GH or HI, which is 40, for a divisor, omitting the cypher to leave place for the next quotient figure, to complete the divisor.

80 into 249 are contained 3 times; this 3 is the width of the oblong AEHG, or HFCI. But the square is imperfect without EBFH; then annex the three to the divisor. Now multiply this perfect divisor by the last figure of the root, to get the quantity in the two oblong figures, and the small square which comprises the great square ABCD.

GHID	=	1600
AEHG	=	120
HFCI	=	120
EBFH	=	9
ABCD	=	1849

How do you find the divisor?

Why do you place the new quotient figure in the units place of the divisor?

How do you prove the square root?

THE CUBE ROOT.

The cube of a number is the product of that number multiplied into its square?

Extraction of the cube root is finding such a number as, being multiplied into its square, will produce the number whose cube root is extracted.

RULE.

Separate the given number into periods of three figures each, beginning at the units place. Find the greatest *cube* in the left hand period, and set its *root* in the quotient; subtract said cube from the period, and to the remainder bring down the next period for a dividual.

Square the root, and multiply the square by *three hundred* for a divisor.

See how often the divisor is contained in the dividual, and place the result in the quotient.

Multiply the divisor by the last found quotient figure; square the last found figure—multiply the square by the preceding figure or figures of the quotient, and this product by *thirty;* and cube the last figure. Add these three products together, and subtract their amount from the dividual.

To the remainder add the next period, and proceed as before, until the periods are all brought down.

When a remainder occurs, annex periods of cyphers to obtain decimals, which may be carried to any convenient number.

NOTE 1.—The cube root of a vulgar fraction is found by reducing it to its lowest terms, and extracting the root of the numerator for a numerator, and of the denominator for a denominator. If it be a surd,* extract the root of its equivalent decimal.

EXAMPLES.

1 What is the cube root of 99252847?

	99,252,847(463	Ans. 463.
$4\times4\times4=64$		
$4\times4\times300=4800$	35252	463
		463
Div. $4800\times6=$	28800	
$6\times6\times4\times30=$	4320	1389
$6\times6\times6=$	216	2778
		1852
Subtrahend	33336	
		214369
$46\times46\times300=634800$	1916847	463
Div. $634800\times3=$	1904400	643107
$3\times3\times46\times30=$	12420	1286214
$3\times3\times3=$	27	857476
Subtrahend	1916847	Proof 99252847

2 What is the cube root of 84604519? Ans. 439.
3 What is the cube root of 259694072? Ans. 638.
4 What is the cube root of 32461759? Ans. 319.
5 What is the cube root of 5735339? Ans. 179.
6 What is the cube root of 48228544? Ans. 364.
7 What is the cube root of 673373097125? Ans. 8765.
8 What is the cube root of 7532641? Ans. 196.02+
9 What is the cube root of 5382674. Ans. 175.2+
10 What is the cube root of 15926.972504?
Ans. 25.16+

When decimals occur, point the periods both ways, beginning at the decimal point, and if the last period of the decimal be not complete, add one or more cyphers.

A mixed number may be reduced to an improper fraction, or a decimal, and the root thereof extracted.

* A surd is a quantity whose root cannot exactly be formed. A quantity whose root *can* be found, is called a *rational* quantity.

1 What is the cube root of $\frac{648}{3000}$? Ans. $\frac{3}{5}$.
2 What is the cube root of $\frac{250}{686}$? Ans $\frac{5}{7}$.
3 What is the cube root of $\frac{1520}{5130}$? Ans $\frac{2}{3}$.
4 What is the cube root of $12\frac{19}{27}$? Ans. $2\frac{1}{3}$.
5 What is the cube root of $31\frac{15}{343}$? Ans. $3\frac{1}{7}$.

SURDS.

6 What is the cube root of $7\frac{1}{5}$? Ans. 1.93+
7 What is the cube root of $9\frac{1}{6}$? Ans. 2.092+

APPLICATION.

1 The cube of a certain number is 103823; what is that number? Ans. 47

2 The cube of a certain number is 1728; what number is it? Ans. 12.

4 There is a cistern or vat of a cubical form, which contains 1331 cubical feet: what are the length, breadth and depth of it? Ans. each 11 feet.

4 A certain stone of a cubical form contains 474552 solid inches; what is the superficial content of one of its sides? Ans. 6084 inches.

Questions.

What is a *cube?* A cube is a solid body contained by six equal square sides.

What is the *cube root?* It is the length of one side of a cube.

What is the *square* of the cube root? It is the superficial contents of one side of a cube.

How do you point off a number whose *cube root* is to be extracted?

What is the first figure of the root? It is the root of the greatest cube in the first period.

When you subtract the cube from the first period, what do you do?

How do you find the divisor?

What is the first step towards finding the subtrahend? What is the second? What is the third?

When a remainder occurs, how do you proceed?

How do you prove the cube root?

Illustration of the Rule for extracting the Cube Root.

The reason for pointing off the number into periods of *three* figures each, is similar to the one given in the Square Root; for the number of figures in any cube will never exceed three times the figures in the root, and will never be more than *two* figures less.

OPERATION.

```
                    15,625 | 25
      2×2×2=         8
                   ------
2×2×300=1200 |     7625
                   ------
                    6000
   5×5×2×30=        1500
      5×5×5=         125
                   ------
                    7625
                   ------
```

In this number there are two periods: of course there will be two figures in the root.

"*The greatest cube in the left hand period* (15) *is* 8, *the root of which is* 2;" therefore, 2 is the first figure of the root, and as we shall have another figure in the root, the 2 stands for 2 tens, or 20. But the cube root is the length of one of the sides of the cube, whose length, breadth and thickness are equal: then the cube whose root is 20, contains 20×20×20=8000.

Fig. 1.

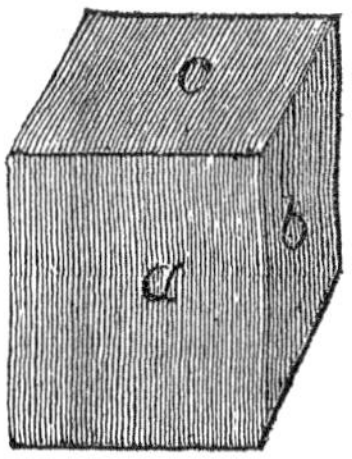

"*Subtract the cube thus found* (8) *from said period, and to the remainder bring down the next period,*" or, subtract the 8000 from the whole given number (15625) and 7625 will remain. Thus 8000 feet are disposed of in the cube, Fig. 1. 20ft long, 80 ft wide, and 20 ft. high.

The cube is to be enlarged by the addition of 7625 feet which remain. In doing this, the figure must be enlarged on *three* sides, to make it *longer*, and *wider*, and *higher*, to maintain the complete cubic form.

The next step is, to find a divisor; and this must be the number of square feet contained in the three sides to which the addition must be made.

Hence we "*multiply the square of the quotient figure by* 300." That is, 2×2×300=1200: or 20×20×3=1200 feet, which is the superficial content of the three sides, A, B, and C.

Fig. 2.

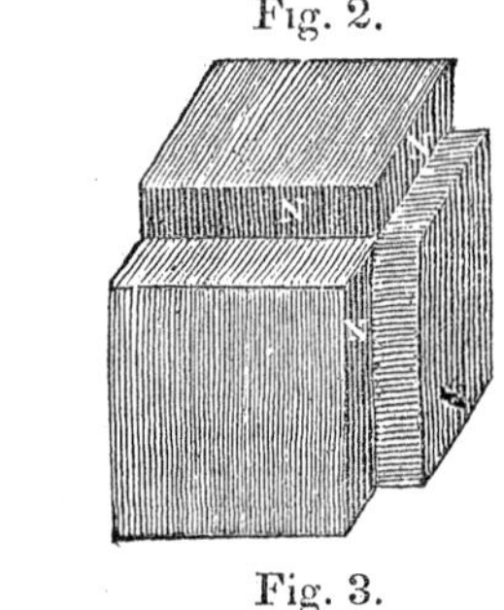

Fig. 3.

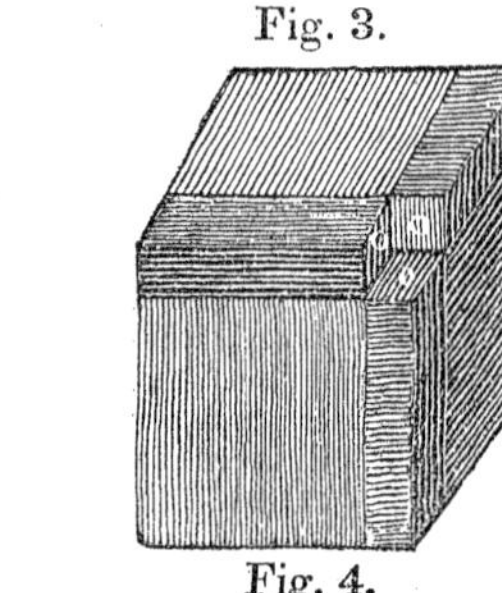

Fig. 4.

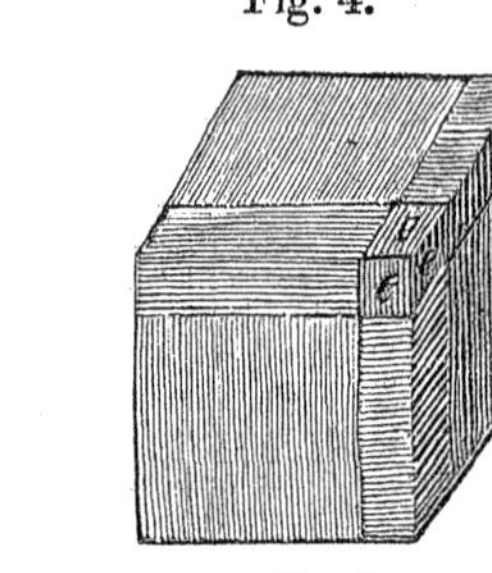

Proof.

20×20×20=	8000
20×20×3×5=	6000
5×5×20×3=	1500
5×5×5=	125
25×25×25=	15625

This "*divisor* (1200) *is contained in the dividual*" (7625) 5 times: then 5 is the second quotient figure; that is, the addition to each of the three sides is 5 feet thick; if 1200 feet cover the three sides one foot thick, 5 feet thick will require 5 times as many; that is 1200×5=6000.

But when the additions are made to the three squares there will be a deficiency along the whole length of the sides of the squares between the additions, which must be supplied before the cube will be complete. These deficiencies will be three, as may be seen at NNN in Fig. 2; therefore it is that we "*multiply the square of the last figure by the preceding figure, and by* 30," (that is, 5×5×2×30,) or 5×5×20×3=1500which is the quantity required to supply the three deficiencies.

Figure 3, represents the solid with these deficiencies supplied, and discovers an other deficiency, where they approach each other at ooo.

Lastly, "*cube the last figure;*" this is done to fill the deficiency left at the corner, in filling up the other deficiencies. This corner is limited by the three portions applied to fill the former vacancies, which were 5 feet in breadth; consequently the *cube* of 5 will be the solid contents of the corner. Fig. 4 represents this deficiency (eee) supplied, and the cube *complete.*

The illustration is much better made by means of 8 blocks of the following description: One cube of about 3 inches diameter; three pieces each 3 inches square, ½ inch thick; three pieces each ½ inch square, 3 inches long; and one cube ½ inch. A set of these should belong to the apparatus of every Professional Teacher.

A GENERAL RULE FOR EXTRACTING THE ROOTS OF ALL POWERS.

1 Point the given number into periods, agreeably to the required root.

2 Find the first figure of the root by the table of powers, or by trial; subtract its power from the left hand period, and to the remainder bring down the first figure in the next period for a dividend.

3 Involve the root to the next inferior power to that which is given, and multiply it by the number denoting the given power, for a divisor; by which find a second figure of the root.

4 Involve the whole ascertained root to the given power, and subtract it from the first and second periods. Bring down the first figure of the next period to the remainder, for a new dividend; to which, find a new divisor, as before; and so proceed.

Note.—The roots of the 4th, 6th, 8th, 9th, and 12th powers, may be obtained more readily thus:

For the 4th root take the square root of the square root.

For the 6th, take the square root of the cube root.

For the 8th, take the square root of the 4th root.

For the 9th, take the cube root of the cube root.

For the 12th, take the cube root of the 4th root.

EXAMPLES.

1 What is the 5th root of 916132832?

```
      9161,32832(62 Ans.
      7776               6×6×6×6×6=7776
      ----               6×6×6×6×5=6480 div.
 6480)13853
      -----
      916132832  62×62×62×62×62=916132832
      916132832
```

2 What is the fourth root of 140283207936? Ans. 612.
3 What is the sixth root of 782757789696? Ans. 96.
4 What is the seventh root of 194754273881? Ans. 41.
5 What is the ninth root of 1352605460594688?
Ans. 48.

ARITHMETICAL PROGRESSION.

A SERIES of numbers, increasing or decreasing by a common difference, is called an *Arithmetical Progression.*

Thus 3, 5, 7, 9, 11, 13, 15, &c., is an ascending series, whose common difference is 2.

And 16, 13, 10, 7, 4, 1, is a descending series, whose common difference is 3.

The three most important properties of an arithmetical series are the following:

I. The sum of the two extremes is equal to *twice* the mean, or to the sum of any two terms equidistant from the mean.

In the above series 3+15= *twice* 9, which is the mean or middle term; and 5+13 which are equidistant from the mean.

II. The difference of the extremes, is equal to the common difference multiplied by the number of terms, less one.

In the above, the number of terms 7—1=6; then the common difference 2×6=12, which is equal to 15—3. Or the number of terms in the other series 6—1=5; then 5×3=15, which is equal to 16—1.

III. The sum of all the terms is equal to the product of the mean, or of half the sum of the extremes, multiplied by the number of terms.

As above, the mean is 9; which, multiplied by 7, the number of terms, gives 63=15+13+11+9+7+5+3.

Or, 15+3=18; half of which is 9. Then, 9×7 =63, as before.

And 16+1=17; half of which is $8\frac{1}{2}$. This multiplied by 6, the number of terms, =51=16+13+10 +7+4+1.

NOTE 1.—*To find the last term*, multiply the common difference by the number of terms, less one, and add the product to the first term in an increasing series; or, subtract the product from the first term in a decreasing series.

EXAMPLES.

1 If the first term is 3, the common difference 2, and the number of terms 7, what is the last term? Ans. 15.

7—1=6; the 2×6+3=15, the last term.

2 The first term being 16, the common difference 3, and the number of terms 6, what is the last term? Ans. 1.

6—1=5; then 3×5=15. Then 16—15=1, the last term.

3 What is the last term in a series, whose first term is 5, the common difference 4, and the number of terms 25? Ans. 101.

4 Suppose, in the above, the first term is 3; what is the last term? Ans. 99.

5 A man bought 50 yards of calico at 6 cents for the first yard, 9 for the second, 12 for the third, &c.; what did he pay for the last? Ans. $1.53.

NOTE 2.—*To find the mean term*, take half the sum of the extremes.

EXAMPLES.

1 The first term is 3, and the last 15; what is the arithmetical mean? Ans. 9.

3+15=18; the 18÷2=9, the mean term.

2 The weight of 5 packages of goods is, severally, 180, 150, 120, 90, 60 pounds; what is the mean or average weight? Ans. 120lbs.

NOTE 3.—*To find the sum of all the terms*, multiply the mean term by the number of terms; or, the mean by the sum of the two extremes, and take half the product.

EXAMPLES.

1 The mean term is 11, and the number of terms 9; what is the sum of the series? Ans. 99.

2 The first term is 5, the last 32, and the number of terms 10; what is the sum of the series? Ans. 185.

3 How many strokes does the hammer of a common clock strike in 12 hours? Ans. 78.

4 What debt can be discharged in one year, by weekly payments in arithmetical progression, the first being $12, and the last, or fifty-second, payment $1236? Ans. 32448.

NOTE 4.—*To find the common difference*, divide the difference of the extremes by the number of terms, less one.

EXAMPLES.

1 The ages of 8 boys form an arithmetical series—the youngest is 4 years old and the oldest is 18; what is the common difference? Ans. 2.

2 A debt can be discharged in one year, by weekly payments in arithmetical progression—the first is $12, and the last $1236; what is the common difference? Ans. $24.

NOTE 5.—*To find the number of terms*, divide the difference of the extremes by the common difference, and add 1 to the quotient.

EXAMPLES.

1 In a series, whose extremes are 4 and 1000, and the common difference 12, what is the number of terms? Ans. 84.

2 If a man, on a journey, travels 18 miles the first day, increasing the distance 2 miles each day, and on the last day goes 48 miles, how many days did he travel? Ans. 16.

PROMISCUOUS EXERCISES.

1 24 persons bestowed charity to a beggar—the first gave him 12 cents, the second 18, &c., in an arithmetical series; what sum did he receive? Ans. $19.44.

2 Suppose 100 apples were placed in a right line, 2 yards apart, and a basket 2 yards from the first; how far would a boy travel to gather them up singly, and return with each separately to the basket?

Ans. 20200 yards.

3 In a drove of 400 hogs, 5 of the largest weighs, on an average, 280 pounds a-piece, and 5 of the smallest 180; what is the mean or average weight of them all, and what is the whole weight?

Ans. { 230lbs. mean weight.
92000 lbs. whole wt. }

4 How many acres in a piece of land 80 rods wide at one end and 60 at the other, and 120 rods long?

Ans. 52½.

It may be observed, that the natural numbers 1, 2, 3, 4, 5, 6, 7, &c., is an arithmetical series, whose first term is 1, and common difference 1; and that the last term is equal to the number of terms.

From this series, we may form another by *adding to each figure the sum of all the preceding*, and we shall have 1, 3, 6, 10, 15, 21, 28, &c.

These are called *triangular numbers*, because they may be represented by points, forming equilateral triangles, thus:

```
                                         .
                          .            .   .
          .             .   .        .   .   .
.       .   .         .   .   .    .   .   .   .
```

Hence we perceive, that the sum of the natural numbers, to any degree, expresses the triangular number of the same degree.

In the same manner, the square numbers 1, 4, 9, 16,

25, 36, &c., may be expressed by points, arranged in squares, thus:

```
                          . . . .
                  . . .   . . . .
          . .     . . .   . . . .
   .      . .     . . .   . . . .
```

Natural numbers—1, 2, 3, 4, 5, 6, 7, 8, 9, &c.

Triangular numbers—1, 3, 6, 10, 15, 21, 28, 36, &c.

Square numbers—1, 4, 9, 16, 25, 36, 49, 64, 81, &c.

Cube numbers—1, 8, 27, 64, 125, 216, 343, 512, 729, &c.

GEOMETRICAL PROGRESSION.

A SERIES of numbers, increasing or decreasing by a common *ratio*, is called a *Geometrical Progression.*

Thus 2, 4, 8, 16, 32, 64, 128, is an increasing series, whose common ratio is 2;

And 729, 243, 81, 27, 9, 3, is a decreasing series, whose common ratio is $\frac{1}{3}$.

The most important properties of a geometrical series, are the following:

I. The product of the extremes is equal to the square of the mean; or, to the product of any two terms equidistant from the mean.

In the above, $128\times2=16\times16$, or 32×8, &c. Also, $729\times3=243\times9$, or 81×27, between which the mean falls.

Hence, the mean term, in a geometrical series, is the *square root* of the product of the extremes, or of any two terms equidistant from the mean.

II. The last term of an increasing series, is the product of the first term, multiplied by the ratio involved to the power which is one less than the number of terms; in a decreasing series, it is the quotient of the first term divided by the power.

In an increasing series, whose first term is 2, ratio 3,

and the number of terms 5, we have, by the natural method, 2, 6, 18, 54, 162, which gives 162 for the last term.

Or, by the artificial method, the ratio involved in the 4th power, which is one less than the number of terms, 3×3×3×3=81; these, multiplied by the first term, 2×81, gives 162, the last term, as before.

Again, let the first term be 4; then we have 4, 12, 36, 108, 324, or 3×3×3×3=81. Then multiply the first term, 4, by 81=324, the last term.

In a decreasing series, with the first term 243, ratio $\frac{1}{3}$, and the number of terms 5, we have 243, 81, 27, 9, 3; then 3×3×3×3=81, and 243÷81 gives 3, which is the last term, as before.

NOTE 1.—*To find the last term*, involve the ratio to the power which is one less than the number of terms, and multiply the first term by the power.

EXAMPLES.

1 What is the eighth term of an increasing geometrical series, whose first term is 4, and ratio 2?

Ans. 512.

2×2×2×2×2×2×2 = 128; then 4×128 = 512, which is the eighth term. Or, 4, 8, 16, 32, 64, 128, 256, 512, the eighth term, as before.

2 Required the last term of an increasing series, whose first term is 15, ratio 2, and number of terms 10.

Ans. 7680.

3 A boy purchased 18 oranges, at 1 cent for the first, 4 for the second, 16 for the third, &c.; what was the price of the last? Ans. $171798691.84.

NOTE 2.—*To find the sum of all the terms*, multiply the last term by the ratio, and from the product subtract the first term; then divide the remainder by the ratio, less one.

EXAMPLES.

1 What is the sum of a series, whose first term is 2, ratio 3, and number of terms 5? Ans. 242.

Find the last term by Note 1.

2	162	last term.
6	3	ratio.
18	486	
54	2	first term.
162		
	3—1=2)484	
242	242	sum of the sums.

2 What is the sum of the series, whose first term is 4, ratio 5, and number of terms 7? Ans. 15624.

Find the last term by Note 1.

4	12500	last term.
20	5	ratio.
100	62500	
500	4	first term.
2500		
12500	5—1=4)62496	
15624	15624	sum of the series.

3 What is the sum of a series, whose first term is 3, ratio 4, and number of terms 7? Ans. 16383.

Last term, ...	12288
Ratio, ...	4
	49152
First term, ...	3
	4—1=3)49149
Sum of the series	16383

Write down the series, multiply it by the ratio, and subtract the first series from the second, thus:

3	12	192	768	3072	12288	
	12	192	768	3072	12288	49152

Here the terms all cancel, but the first of the upper and last of the lower series. Then we have

	49152
	3
Divide by ratio, less 1: 4—1=3)	49149
The sum of the series, ...	16383

Now, as we multiplied the given series by the ratio, which is 4, and subtracted *once* the series from the product, the remainder is *three times the given series.* We, therefore, divide by 3, which is the ratio, less 1 : the quotient is the sum of the series.

4 At 2 cents for the first ounce, 6 for the second, 18 for the third, &c., what would a pound of gold cost? Ans. \$5314.40.

5 Sold 10 yards of velvet, at 4 mills for the first yard, 20 for the second, 100 for the third, &c.; what did the piece cost? Ans. \$9765.62,4.

6 What is the cost of a coat with 14 buttons, at 5 mills for the first, 15 for the second, 45 for the third, &c.? Ans. \$11957.42.

7 What is the cost of 16 yards of cloth, at 3 cents for the first yard, 12 for the second, 48 for the third, &c.? Ans. \$42949672.95.

NOTE 3.—The *sum* of a geometrical series is found by the extremes and the ratio, independent of the number of terms; hence, whether the number of terms be *many* or *few*, there is no variation in the rule. We may, therefore, require the sum of the seres, 6, 3, 1, $\frac{1}{3}$, $\frac{1}{9}$, &c., to infinity, provided we can determine the value of the other extreme. Now, we see the terms decrease as the series advances; and the hundredth term, for example, would be exceedingly small, the thousandth too small to be estimated, the millionth still less, and the infinite term would be nothing: not, as some tell us, "extremely small," or, "too little to be considered," &c., but *absolutely nothing*.

Now let us consider the series as inverted; then 6 will be the last term, and 3 the ratio. By the rule, *multiply the last term by the ratio, subtract the first term, and divide by the ratio, less* 1. Here we have the sum of the series, $\frac{6\times 3-0}{2}=9$, the answer.

EXAMPLES.

1 What is the sum of the infinite series, $\frac{1}{2}+\frac{1}{4}+\frac{1}{8}+\frac{1}{16}$, &c.? Invert the series: $\frac{1}{2}$ is the last term, and 2 the ratio; hence, $\frac{\frac{1}{2}\times 10-0}{1}=1$, the answer.

2 What is the sum of the infinite series $\frac{3}{10}$, $\frac{3}{100}$, $\frac{3}{1000}$, &c.? $\frac{\frac{3}{10}\times 10-0}{9}=\frac{1}{3}$, the answer.

3 What is the value of $\frac{1}{5}$, $\frac{1}{25}$, $\frac{1}{125}$, &c., to infinity? Ans. $\frac{1}{4}$.

4 What is the value of $\frac{3}{4}$, $\frac{1}{2}$, $\frac{1}{3}$, $\frac{2}{9}$, &c., to infinity? Ans. $\frac{9}{4}$.

5 What is the value of 1, $\frac{3}{4}$, $\frac{9}{16}$, &c., to infinity? Ans. 4.

Here the ratio is $\frac{4}{3}$.

6 What is the value of $\frac{2}{5}$, $\frac{4}{25}$, $\frac{8}{125}$, &c., to infinity? Ans. $\frac{2}{3}$.

7 What is the value of .777, &c., to infinity? This may be expressed by $\frac{7}{10}$, $\frac{7}{100}$, $\frac{7}{1000}$, &c. Ans. $\frac{7}{9}$.

8 What is the sum of .6666, &c., to infinity? Ans. $\frac{2}{3}$.

9 What is the value of .232323, &c., infinitely extended? Ans. $\frac{23}{99}$.

This may be expressed by $\frac{23}{100}$, $\frac{23}{1000}$, &c.

EXCHANGE.

THE object of exchange is to find how much of the money of one country is equivalent to a given sum of the money of another.

By the *par of exchange* between two countries, is meant the intrinsic value of the one, compared with the other; it is estimated by the weight and fineness of the coins.

The *course of exchange*, at any time, is the sum of the money of one country which, at that time, is given for a certain sum of money of another country. The course of exchange varies according to the circumstances of trade. All the calculations in exchange can be performed by the Rule of Three.

EXAMPLES.

1 A sovereign, of England, is worth $4.84,6. A merchant of New York is indebted to a merchant of London $7520; how many sovereigns will it require to discharge the debt?

2 A six-ducat piece of Naples, is worth $5.25. A merchant of Naples is indebted to a merchant in Boston $6940; how many such pieces shall he remit?

3 The current rupee of Calcutta is 44.4 cents. A merchant of Philadelphia has a claim against a merchant of Calcutta of $437; how many rupees shall he draw?

4 The piaster is $20\frac{1}{2}$ cents; how many piasters in $1128.22?

5 Reduce 7218 rupees to Federal money, at 46 cents per rupee.

6 The dollar of Bencoolen is reckoned at $1.10, Federal money; how much Federal money in $2740 of Bencoolen?

7 The Prussian rix-dollar is worth $66\frac{2}{3}$ cents. Reduce $7348.32 into Prussian money.

Ex. 1. If 1£ sterling is worth \$4,44 cts. 4 ms.; what is 65£ sterling worth? Ans. \$288,86.

2. What is the value of \$500 in English money, at \$4,44,4, per £ sterling? Ans. 112£ 10s. $2\frac{1}{2}$d.

3. What is the value of 125£ 7s. at \$4,44,4, per £ sterling? Ans. \$557,05,5.

4. What is the value of \$1000, in English money, at \$4,44,4, per £ sterling Ans. 225£ $5\frac{1}{4}$d.

PROMISCUOUS EXERCISES.

1. A merchant had 1000 dollars in bank; he drew out at one time \$237.50, at another time, \$116.09, and at another, \$241.06½: after which he deposited at one time 1500 dollars, and at another time \$750.50; how much had he in bank after making the last deposite? Ans. \$2655.84½.

2 Sold 8 bales of linen, 4 of which contained 9 pieces each, and in each piece was 35 yards; the other 4 bales contained 12 pieces each, and in each piece was 27 yards; how many pieces and how many yards were in all? Ans. 84 pieces, 2556 yards.

3 If a man leave 6509 dollars to his wife and two sons, thus—to his wife $\frac{3}{8}$, to his elder son $\frac{3}{5}$ of the remainder, and to his other son the rest; what is the share of each?

Ans. { Wife's share \$2440.87½.
Elder son's \$2440.87½.
Other son's \$1627.25. }

4 What is the commission on \$2176.50, at $2\frac{1}{2}$ per cent? Ans. \$54.41¼.

5 If a tower is 384 feet high from the foundation, a sixth part of which is under the earth, and an eighth part under water, how much is visible above the water? Ans. 272 feet.

6 How many bricks 9 inches long and 4 inches wide, will pave a yard that is 20 feet square? Ans. 1600.

7 What is the value of a slab of marble, the length of which is 5 feet 7 inches, and the breadth 1 foot 10 inches, at 1 dollar per foot? Ans. \$10.23½.

8 A certain stone measures 4 feet 6 inches in length,

2 feet 9 inches in breadth, and 3 feet 4 inches in depth; how many solid feet does it contain? Ans. 41 ft. 3 in.

9 A line 35 yards long will exactly reach from the top of a fort, standing on the brink of a river, to the opposite bank, known to be 27 yards from the foot of the wall; what is the height of the wall?

Ans. 22 yards $3\frac{1}{4}$ inches.+

10 The account of a certain school is as follows, viz. $\frac{1}{16}$ of the boys learn geometry, $\frac{3}{8}$ learn grammar, $\frac{3}{10}$ learn arithmetic, $\frac{3}{20}$ learn to write, and 9 learn to read: what number is there of each?

Ans. { 5 who learn geometry, 30 grammar, 24 arithmetic, 12 writing, and 9 reading.

11 A merchant, in bartering with a farmer for wood at \$5 per cord, rated his molasses at \$25 per hhd., which was worth no more than \$20; what price ought the farmer to have asked for his wood to be equal to the merchant's bartering price? Ans. \$6,25.

12 A and B dissolve partnership, and equally divide their gain: A's share, which was \$332 50 cts., lay for 21 months; B's for 9 months only: the adventure of B. is required. Ans. \$775 83⅓ cts.

13 If a water-hogshead holds 110 gals. and the pipe which fills the hogshead discharges 15 gal. in 3 minutes, and the tap will discharge 20 gal. in 5 minutes, and these were both left running one hour, how many gallons would the hogshead then contain; and if the tap was then stopped, in what time would the hogshead be filled?

Ans. 60 gal., and filled in 10 min.

14 A has B's note for \$500 75 cts., with 9 months interest, at 6 per cent., due on it, for which B gave him 5064 feet of boards, at 2½ cents per foot, with 140 pounds of tallow, at 13 cts. per pound, and is to pay the rest in flax seed, at 92½ cts. per bushel; how many bushels of flax seed must A receive, to balance the note?

Ans. $409\frac{159}{925}$. bushels.

15 A, B, and C, in company, had put in \$5762: A's money was in 5 months, B's 7, and C's 9 months: they gained \$780, which was so divided, that $\frac{1}{4}$ of A's was $\frac{1}{5}$ of B's, and $\frac{1}{5}$ of B's was $\frac{1}{3}$ of C's: but B, having received

$2087, absconded: what did each gain, and put in; and what did A and C gain or lose by B's misconduct?

Ans.	A's stock	$2494,887	gain 260
	B's do	$2227,577	do 325
	C's do	$1039,536	do 195
	A and C would gain		$465,577

16 When 100 boxes of prunes cost 2 dollars 10 cents each, and by selling them at 3 dollars 50 cents per cwt. the gain is 25 per cent., the weight of each box, one with another, is required. Ans. 84 lb.

17 There are two columns, in the ruins of Persepolis, left standing upright; one is 64 feet above the plain, the other 50. Between these, in a right line stands an ancient statue, the head whereof is 97 feet from the summit of the higher, and 86 feet from the top of the lower column, and the distance between the lower column and the centre of the statue's base, is 76 feet; the distance between the top of the columns is required. Ans. 157+ft.

18 If I see the flash of a cannon, fired from a fort on the other side of a river, and hear the report 47 seconds afterwards, what distance was the fort from where I stood? Ans. 53674 feet.

NOTE.—Sound, if not interrupted, will move at the rate of about 1142 feet in a second of time.

19 What is the difference between the interest of $1000 at 6 per cent. for 8 years, and the discount of the same sum at the same rate, and for the same time?

Ans. The interest exceeds the discount by $155 67 cts. 5 m.

20 If a tower be built in the following manner, $\frac{87}{170}$ of its height of stone, 27 feet of brick, and $\frac{1}{4}$ of its height of wood, what was the height of the tower?

Ans. 113 feet 4 inches.

21 A captain, 2 lieutenants, and 30 seamen, take a prize worth $7002, which they divide into 100 shares, of which the captain takes 12, the two lieutenants each 5, and the remainder is to be divided equally among the sailors; how much will each man receive? Ans. Captain's share, $840,24, each lieutenant's, $350,10, and each seaman's, $182,05,2.

PREFACE TO APPENDIX.

The author of the foregoing work, has long contemplated its extension in an Appendix, which is now offered to the public in hopes the usefulness of the whole may be extended, and the science of arithmetic advanced. The view of numbers, and the abridged modes of operation herein presented, it is believed, will be found acceptable alike to the business-man and to the scholar.

The Appendix recognizes the scientific character of numbers, and gives bold and enlarged views of arithmetical operations. The method of *cancelling* is not new: but for a long period it has scarcely been known. It is, however, coming into general use; and it is carried much farther in this part of the work than in any we have hitherto known.

In Europe, this system has been very generally adopted in the higher schools, and in this country it is fast becoming known—and as far as it is known, it supercedes the usual modes of operation.

To the method of stating problems in Proportion by comparing *cause* and *effect*, we invite special attention. On critical examination it will be found more easy and more rational than any other method.

Other methods of statement sometimes require the number of men to be multiplied into feet of wall—days into acres of grass, &c., all of which, though correct as *abstract proportion in numbers*, is unnatural and void of strict philosophical expression—not so with this method.

The peculiarities which the student will here find in the Extraction of Roots, and in Mensuration, are not *all* new—indeed there can be nothing new in *principle*—but as far as the author's knowledge extends, he is not aware that these abbreviations have ever been collected in any arithmetical work. The impression seems to have been, that the people could not comprehend arithmetical brevity, nor appreciate mathematical beauty; but the author thinking otherwise, presents this brief, yet comprehensive, Appendix to the public, in the full assurance that whoever will pay due attention to the subject will be highly gratified and abundantly rewarded.

APPENDIX.

THE pupil having passed over the common routine of Arithmetic, and supposed to be able to perform all its necessary operations in the usual way, we now present him with some new modes of practical operations, by which the labors will be much abridged and the science presented with more of its roses and less of its thorns.

We commence with a systematic study of numbers. The following are called *prime* numbers; because no one can be divided by any number less than itself without producing a fraction. 1 2 3 . 5 . 7 . . . 11 . 13 . . . 17 . 19 . . . 23 . . . . . 29 . 31 . . . . . 37 . . . 41 . 43 . . . 47 . . . . . 53 . . . 57 . 59 . 61 . . . . 67 . . . 71 . 73 . . . . . 79 . . . 83 . . . 87 . 89 . . . . 97 . . . 101

The points represent the composite numbers; and here it can be observed that there are 29 *prime* numbers, and of course 71 composite numbers in the first hundred—the prime numbers becoming fewer as the numbers rise higher. Observe the following series:

5 10 15 20 25 30 35 40 45 50,

and so on. Every body knows that our Geometric scale of numbers is 1 10 100 1000, &c. Now we wish the student to observe the numbers,

5 20 25 50 75 125 500,

as being not only in the preceding series, but *aliquot parts* of some number in our Geometric scale. For example, 25 is $\frac{1}{4}$ of 100; 125 is $\frac{1}{8}$ of 1000, &c.

We now charge the student to make his eye familiar with all the preceding series—the prime numbers as being unmanageable and inconvenient, and the others the very reverse; but the full importance of such a study can only appear in the sequel.

By a little attention to the relation of numbers, we may often contract operations in multiplication. A dead unifomity of operation *in all cases* indicates a mechanical and not a scientific knowledge of numbers. As a uniform principle, it is much easier to multiply by the small numbers 2, 3, 4, 5, than by 7, 8, 9.

EXAMPLES.

Multiply 4532
by 39

Commence with the 3 tens. Multiply this 13596 by 3, because $3\times3=9$, and place the product in the place of units.

```
 13596
40788
------
176748
```

```
Multiply . .    576
by . . . .      186
             ------
(6×3=18.)      3456
              10368
             ------
             107136
```

Multiply this last number, 3456, (which is 6 times 576) by 3, and place the product in the place of tens, and we have 180 times 576. Observe the same principle in the following examples.

```
Multiply . . . .    576
by . . . . . .      618
                 ------
Commence with 6.   3456
(6×3=18.)         10368
                 ------
                 355968
```

```
Multiply . .      40788
by . . . .          497
                -------
(7×7=49.)        285516
                1998612
                -------
               20271636
```

Observe, that in this last example 497 is 3 less than 500; 500 is $\frac{1}{2}$ of 1000, thefore

```
                               2 ) 4 0 7 8 8 ' 0 0 0
                                   -----------------
                                   2 0 3 9 4   0 0 0
Subtract (3×40788=122364.)           1 2 2   3 6 4
                                   -----------------
                                   2 0 2 7 1   6 3 6
```

Multiply	7 8 5 4 6 0
by	1 4 4 1 2

First multiply by 12, then that product by 12.

```
         9 4 2 5 5 2 0
   1 1 3 1 0 6 2 4 0
   -------------------
   1 1 3 2 0 0 4 9 5 2 0
```

Multiply 86416 by 135. Observe, 135 is 125+10 125 is $\frac{1}{8}$ of 1000, therefore

```
8)8 6 4 1 6'0 0 0
  ---------------
   1 0 8 0 2 0 0 0
       8 6 4 1 6 0
  ---------------
Product . . . . . 1 1 6 6 6 1 6 0
```

The following are from Ray's Arithmetic, page 34:

Multiply 1646 by 365. As the first factor is *even*, and the last ends in 5, we mentally *half* the one and *double* the other, which will not affect the product, but very much contract the operation, we then have

```
  8 2 3
    7 3 0
  ---------
  2 4 6 9 0
5 7 6 1
---------
6 0 0 7 9 0
```

We can do the same in all such cases.

Multiply 999 by 777.

```
                7 7 7 0 0 0
Subtract . . . .      7 7 7
                -----------
Product . . . . 7 7 6 2 2 3
```

Multiply . .	61524
by	7209

	553716	Multiply this product of 9 by
	4429728	8, because 9 times 8 are 72, and place the product in the place
Product	443646516	of 100, because it is 7200.

Multiply	1243	
by	636	
	7458	First by 600.
	44748	Multiply 7458 by 6.
Product	790548	

Multiply 624 by 85. The product will be the same as the half of 624 by the double of 85—that is, 312 by 170, or 3120 by 17=53040. We do not say that these changes give any advantage in *this particular example;* we only make them to call out thought and attention from the pupil.

Multiply	7 8 6 4
by	2 4 6

This may be done by commencing with the 2; then that product by 2 and 3. Or we may commence with the 6 units, and then that product by 4; because 4 times 6 are 24.

Multiply 4386, or any other number, by 49. We may do this by taking the number 50 times, or $\frac{1}{2}$ of 100 times, and subtracting once the number.

Multiply 87742 by 65. This *may be done* by taking the number $\frac{1}{2}$ (100,) +10+$\frac{1}{2}$ of (10) times. That is

	4 3 8 7 1 0 0
	8 7 7 4 2 0
	4 3 8 7 1 0
Product	5 7 0 3 2 3 0

Multiply 92636 by 150. It will be much more easy to multiply the half of it by 300, which will give the same result.

Multiply 679 by 279. Multiply first by 9, and that product by 3, put in the place of 10.

Multiply 87603 by 9865. By the common formal rule this would be a tedious operation; but let us observe that 9865 is 10000—135, 135 is 125+10; 125 is $\frac{1}{8}$ of 1000. Therefore,

```
876030000            Subtract 1/8, 8)87603000
 11826405                        ----------
---------                         10950375
864203595 Product.      10 times,   776030
                                  --------
                                  11826405
```

Multiply 818327 by 9874. But 9874 is 10000 less 126. Therefore

```
          8183270000      Less 1/8 of 818327000
Subtract   103108202                  ---------
          ----------                  102290875
          8080161798 Product.            818327
                                      ---------
                                      103109202
```

Multiply 188 by 135, and that product by 15. Instead of doing this literally and mechanically according to rule, we may half 188 twice, and double each of the factors that end in 5, and we shall have 47 . 270 . . 30 ; or,

```
   4 7
   8 1 0 0
 ---------
   4 7 0 0
 3 7 6
 ---------
 3 8 0 7 0 0
```

How far will a ship sail in 365 days, at the rate

of 8 miles per hour? Here $365\times24\times8=730\times96$; or,

$$\begin{array}{lr} & 73000 \text{ less } 730\times4=2920 \\ & 2920 \\ \hline \text{Product} \ldots\ldots & 70080 \end{array}$$

EXERCISES FOR PRACTICE.

$299\times299\times299=$what number? Ans. 26730899

$999\times999\times999=$what number? Ans. 997002999

$4962\times98=$what number? Ans. 487276

Multiply 8340745 by 64324. Observe, that 32 is 8 times 4, and 64 is 2 times 32.

Multiply 8340745 by 64432. In this last example, commence at the 4 in the place of hundreds.

Multiply $24\times25\times12\times5$, together. Here it would be no index of even a decent knowledge of numbers, to obey literally and in the order the numbers are given; yet how many even at the present day would do so!

Take the factor 4, *out of the* 24, and multiply it into the 25; also, the factor 2, out of 12, and put it with 5, which can be done without effort. We then have $6\times100\times10\times6=36000$, the answer.

SECTION II.

We shall say nothing of division in whole numbers, as nothing new or interesting can be offered on that topic; but we cannot forbear making a few comments on division in decimals.

To divide, to cut into parts, will not at all times give a clear understanding of the operation, and confusion frequently arises from taking this view of the subject; we better consider it as one number *measuring* another. For example: how often will .5 of a foot measure 12 feet? In other words, divide 12 by .5; or, divide 12 by

$\frac{1}{2}$. Here, if the student should imagine that 12 must be cut into parts, he would make a great error. He must divide 120 tenths into parts; in this case, into 5 parts— because the 5 is .5: or he may consider that $\frac{1}{2}$ of a foot may be laid down in 12 feet; that is, measure 12 feet 24 times. Or he may reduce the 12 feet to half feet, and then divide by 1. In all cases, the divisor and dividend must be of the same denomination before the division can be effected. But in decimals, these reductions are made so easily, that a *thoughtless* operator rarely perceives them; hence the difficulty in ascertaining the value of the quotient.

We now give a few examples, for the purpose of teaching the pupil how to use his judgment; he will then have learned a rule *more valuable* than all others.

EXAMPLES.

Divide 15.34 by 2.7. Here we consider the whole number, 15, is to be divided by less than 3; the quotient must, therefore, be a little over 5. One figure then, in the quotient, will be whole numbers, the rest decimals.

Divide 15.34 by .27. Here we perceive that 15 is to be divided, *or rather measured*, by less then $\frac{1}{3}$ of 1; therefore the quotient must be more than 3 times 15. Or we may multiply both dividend and divisor by 100, which will not effect the quotient, and then we shall have 1534, to be divided by 27. Now no one can mistake how much of the quotient will be whole numbers: the rest, of course, decimals.

Divide 45.30 by .015. Conceive both numbers to be multiplied by 1000; then the requirement will be to divide 45300 by 15, a common example in whole numbers.

By attention to this operation, the student will have no difficulty in any case where the divisor is *less* than the dividend.

Here is one of the most difficult cases:

Divide .003753 by 625.5. In all such examples as this, we *insist* on the formality of placing a cipher in the dividend, to represent the place of whole numbers, thus:

625.5)0.0037530(

We now consider whether the whole number in the divisor will be contained in the whole number in the dividend, and we find it will not; we, therefore, write a *cipher* in the quotient to represent the place of whole numbers, and make the decimal on its right, thus, 0.

We now consider, that 625 will not go in the 10's, nor in the 100's, nor in the 3, nor in the 37, nor in the 375, but it will go in the 3753.

We must make a trial at every step, that is, every time we take *in view another place;* and we must take *but one at a time.* In this case, then, we shall have 0.000006, the quotient.

Divide 3 by 30. 30 will not go in 3; we, therefore, write 0 for place of whole numbers, and then say 30 in 30 tenths, 1 tenth times; or, 0.1.

Divide .55 by 11.

11)0.55(0.05

11 in 0, no times; 11 in 5 tenths, 0 times; 11 in 55 hundredths, 5 hundredths times.

It will be observed, that we make the decimal point in the quotient as soon as we ascertain it; not wait, and then find where it should be by counting, &c.—a rule that we regard as unworthy of being followed by all those who can use their reason.

EXAMPLES.

1 Divide 9 by 450. Ans. 0.02.

2 Divide 2.39015 by .007. Anss. 341.45.

3 Divide 100 by .25. Ans. 400.00.

4 If 350 pounds of beef cost $12.25, what is the cost of one pound? Ans. .035.

5 At $5.75 per yard, how much cloth can be purchased with $19.50625? Ans. 3.375 yards.

6 At .07 per cent. per annum, how much capital must be invested to yield $602 ? Ans. $8600.

7 A benevolent individual gave away $600 per annum to charitable objects, which was .12 of his income. What was his income ?

SECTION III.

Multiplication and Division Combined.

WHEN it becomes necessary to multiply two or more numbers together, and divide by a third, or by a product of a third and fourth, it must be *literally done, if the numbers* are *prime.*

For example: Multiply 19 by 13, and divide that product by 7.

This must be done at full length, because the numbers are *prime;* and in all such cases there will result a *fraction.*

But when two or more of the numbers are *composite numbers*, the work *can always* be contracted.

Example: Multiply 375 by 7, and divide that product by 21. To obtain the answer, it is sufficient to divide 376 by 3, which gives 125.

The 7 divides the 21, and the factor 3 remains for a divisor.

Here it becomes necessary to lay down a *plan of operation.*

Draw a perpendicular line, and place all numbers that are to be multiplied together under each other, on the right hand side, and all numbers that are divisors under each other, on the left hand side.

EXAMPLES.

1 Multiply 140 by 36, and divide that product by 84. We place the numbers thus:

84	140
	36

We may cast out *equal factors* from each side of the line *without affecting the result.* In this case, 12 will divide 84 and 36. Then the numbers will stand thus:

7	140
	3

But 7 divides 140, and gives 20, which, multiplied by 3, gives 60 for the result.

2 Multiply 4783 by 39, and divide that product by 13:

$\not{1}\not{3}$	4783
	$\not{3}\not{9}$ 3.

Three times 4783 must be the result.

3. Multiply 80 by 9, that product by 21, and divide the whole by the product of 60×6×14.

3 $\not{6}\not{0}$	$\not{8}\not{0}$ 4
6	9
2 $\not{1}\not{4}$	$\not{2}\not{1}$ 3

In the above, divide 60 and 80 by 20, and 14 and 21 by 7, and those numbers will stand cancelled as above, with 3 and 4, 2 and 3 at their sides.

Now the product 3×6×2, on the divisor side, is equal to 4 times 9 on the other, and the remaining 3 is the result.

Hoping the reader now understands our forms, and comprehends the true philosophical principle, we will give no more abstract examples; but we will give many practical examples, such as might occur in business, and we prefer taking them from books, that it may not be said we made them expressly for this occasion.

Again, it may be observed, that this method of operation may serve for only a few problems. We answer, it will serve for 71 out of 100, according to the theory of numbers, as we have seen there are 71 composite numbers in the first hundred, and more as they rise higher. But the *prime* numbers, 2, 3, and 5, are so small and manageable, and are factors in so many other numbers, that they may be considered in as favorable a light as the composite numbers—and for this reason we may say, 75 out of every hundred problems can be abridged.

But, in *actual business*, the problems are *almost all* reduceable by short operations; as the prices of articles, or amount called for, always corresponds with some *aliquot* part of our scale of computation.

This method may not work a great many problems as they are found in some books, but it will work 90 out of every 100 that *ought* to be found in books.

In a book, we might find a problem like this:

What is the cost of 2lb. 7oz. 13dwt. of tea at 7s. 5d. per pound? But the person who should go to a store and call for 3lb. 7oz. and 13dwt. of tea, would be a fit subject for a mad-house. The above problem requires downright drudgery, which every one ought to be able to perform, but such drudgery *never occurs in business.*

The following examples are extracted from books in common use, and we mark them in order that any one may find the original. For instance, T. 42, means Talbott's Arithmetic, page 42; R. 93, means Ray's Arithmetic, page 93; E. 123, means Emerson's Arithmetic, page 123, &c.

EXAMPLES.

4 How many bushels of apples can be bought for $3, at 15 cts. a peck? (R. 92.) Ans. 5.

$\cancel{5}\ \cancel{15}\ \big|\ \cancel{3}$
$\cancel{4}\ \big|\ \cancel{100}\ 5$

Explanation—3 in 15, 5 times; 4 times 5 are 20, and 20 in 100, 5 times.

5 A farmer has 91 bushels of wheat, and he wishes to put it into bags, each of which holds 3 bushels 2 pecks; how many bags will it take? (R. 92.) Ans. 26.

3bu. 2 pe.=14pe. $\cancel{7}\ \cancel{14}\ \big|\ \cancel{91}\ 13$
$\big|\ \cancel{4}\ 2$

6 What is the value of a piece of gold weighing 1lb. 3dwt., at $12\frac{1}{2}$cts. per grain? (R. 92.) Ans. $729.

$\big|\ 243$
$\cancel{8}\ \big|\ \cancel{24}\ 3$

7 At 3cts. a pound what will 6cwt. 1qr. cost? (R. 83.) Ans. $21.

	~~25~~
~~100~~	~~28~~ 7
	3

8 At $2.25 a qr. what will 1 ton 1cwt. cost? (R. 93.) Ans. $189.

	21 cwt.
~~4~~	~~4~~
	9

9 At 5cts. per oz., what will 7 lbs. 8oz. cost? (R. 93.) Ans. $6.

10 In this example, we must reduce 7lbs. 8oz. to ounces: $7\times16+8$ is the same as $14\times8+8$, or 15×8.

	15			30
100	8	} =100		2
	5			10

11 A grocer bought a lot of cheese, each weighing 9lbs. 15oz., the weight of the whole amounted to 39cwt. 3qr.; how many cheeses were there? (R. 93.) Ans. 448.

	~~159~~ qr.
~~159~~	28
	16

12 How many casks, each holding 84lbs., can be filled out of a hogshead of sugar weighing 15cwt. 3qr? (R. 93.) Ans. 17.

13 A bell of Moscow weighs 288000lbs.; how many tons? (R. 93—part of Ex. 23.)

Ans. 128 tons. 11cwt. 1qr. 20lbs.

14 We prefer dividing (mentally) the pounds into their obvious factors.

7 ~~28~~	~~2~~
~~4~~	~~144~~ ~~30~~ 9
~~20~~	100~~0~~

7)900

$128\frac{4}{7}$

15 How many times will a wheel, which is 9ft. 2in.

in circumference, turn round in going 65 miles? (R. 94, Ex. 32.) Ans. 37440.

110	65
2	320
	12

16 What will 2 square yards 2 square feet of ground come to at 5cts. a square inch? (R. 94.) Ans. $144

2 square yards 2 square feet=20 square feet.

	~~20~~
	12
~~100~~	12
	~~5~~

17 What will one square yard of gilding cost at 12.5 cents a square inch? (R. 94.) Ans. $162.

8	9
	144

18 What will 5 yards 2qrs. of cloth cost at $12\frac{1}{2}$cts. a nail? (R. 96.) Ans. $11.

8	22 qrs.
	4

19 How many coat patterns, each containing 3 yards 2qrs., can be cut out of a piece of cloth containing 70 Ells Flemish? (R. 95.) Ans. 15.

14	70
	3

20 What will one hhd. of wine cost at $6\frac{1}{4}$cts. a gill? (R. 95.) Ans. $126.

Observe, that $6\frac{1}{4}=\frac{25}{4}$; and, as 4 is a divisor to 25, it must be put on the opposite side of the line.

	63
100	4
4	2
	4
	25

or,

	63
16	4
	2
	4

21 If a person write 10 minutes each day, how

much time will that amount to in 4 years? (R. 96.)
Ans. 10 days $3\frac{1}{2}$ hours.

60 24	365 4 10

22 How many yards of carpeting, 2 feet 6 inches in breadth, will cover a floor 27 feet long and 20 feet wide? (T. 98.) Ans. 72.

$2\frac{1}{2}=\frac{5}{2}$. 2 is to divide the 5; it must, therefore, go over the line.

5 3	20 27 2

23 What quantity of shalloon, 3 quarters wide, will line $7\frac{1}{2}$ yards of cloth that is $1\frac{1}{2}$ yards wide? (T. 98.)
Ans. 15.

$\frac{3}{4}$	$7\frac{1}{2}$ $1\frac{1}{2}$

or,

~~2~~ ~~2~~ ~~3~~	15 ~~3~~ ~~4~~

N. B. Mixed numbers are reduced to improper fractions, and the denominators thrown over the line.

24 How much land, at $2.50 per acre, must be given in exchange for 360 acres, worth $3.75. (T. 99.)
Ans. $540.

$2\frac{1}{2}$	360 $3\frac{3}{4}$

or,

~~4~~ ~~5~~	~~360~~ 90 ~~15~~ 3 2

25. What will a bushel of clover-seed come to at $12\frac{1}{2}$ cts. a pint? (Wilson, 41.) Ans. $8.

$12\frac{1}{2}$ cts.$=\frac{1}{8}$ of a dollar. The 8 on one side cancels 8 on the other, and leaves 4×2 for the answer.

~~8~~	4 pecks. ~~8~~ 2

26 Suppose a hogshead of molasses, which cost $23, be retailed at $12\frac{1}{2}$cts. a quart; what is the profit on it? (W. 41.) Ans. $8.50.

2 8 | 63 gallons. Sale 31.50
| 4 Cost 23

Profit 8.50

27 What will 5 barrels of flour cost at $3\frac{1}{2}$cts. per pound? Ans. \$34.30.

28 How many times is $\frac{2}{3}$ of a pint contained in $\frac{5}{9}$ of a gallon? (W. 65.) Ans. $6\frac{2}{3}$.

$\frac{2}{3}$ | $\frac{5}{9}$ 4 2 or, 3 9 | 5 3 4 2 = 3 | 20 / $6\frac{2}{3}$
2

We have already remarked, that denominators of fractions must go over the line *from the term to which they belong.*

29 How many times can a vessel, holding $\frac{9}{10}$ of a quart, be filled from $\frac{1}{3}$ of a barrel containing $31\frac{1}{2}$ gallons? (W. 66.) Ans. $46\frac{2}{3}$.

3 | $31\frac{1}{2}$ = 2 | 63 7
$\frac{9}{10}$ | 5 3 | 4 2
9 | 10

30 If one acre and 20 rods of ground produce 45 bushels of wheat; at that rate, how much will nine acres produce? (W. 90.) Ans. 360.

1a. 20r.=180. 20 180 | 9
| 160 8 45×8=360.
| 45

N. B. We shall plan no more problems in this section; but the following require no *real labor,* save correct reasoning. When the numbers are properly arranged, a few clips with the pencil, and perhaps a *trifling* multiplication, will suffice.

31 At $1\frac{1}{2}$cts. a gill, how many gallons of cider can be bought for \$12? (R. 95.) Ans. 25.

32 How many casks, each containing 12 gallons, can be filled out of a ton of wine? (R. 95.) Ans. 21.

33 A man retailed 9 barrels of ale, and received for it \$129.60 ; at what price did he sell it a pint? (R. 96.)
Ans. 5cts.

34 How much butter, at 9cts. per pound, will pay for 12 yards of cloth, at \$2.19 per yard? (W. 79.)
Ans. 292.

35 At $45\frac{1}{2}$ dollars per acre, what will 32 rods of land come to? (W. 79.) Ans. \$9.10.

36 How long must a laborer work, at $62\frac{1}{2}$cts. a day, to earn \$25 ? (W. 78.) Ans. 40 days.

37 A merchant sold 275 pounds of iron, at $5\frac{1}{4}$cts. a pound, and took his pay in oats, at 50cts. a bushel; how many bushels did he receive? (Adams, 53.)
Ans. $38\frac{3}{8}$.

38 How many yards of cloth, at \$4.66 a yard, must be given for 18 barrels of flour, at \$9.32 a barrel? (A. 53.) Ans. 36.

39 How long will it require one of the heavenly bodies to move through a quadrant, at the rate of 43′ 12″ per minute? (R. 97.) Ans. $2\frac{1}{12}$ hours.

40 If a comet move through an arc of 7° 12′ per day, how long would it be in passing an arc of 180°? (R. 97.) Ans. 25 days.

41. What is the cost of 8hhds. of wine, at 5cts. per pint? Ans. \$201.60.

42 There are $30\frac{1}{4}$ square yards in one perch of land; how many perches are there in 363 square yards?
Ans. 12.

43 What will $18\frac{3}{4}$ yards cost, at 75cts. per yard?
Ans. \$$14.06\frac{1}{4}$.

44 If 16 persons receive \$516 for 43 days' work, how much does each man earn per day? Ans. 75cts.

45 How many times will a wheel, which is 12 feet in circumference, turn round in going a mile? Ans. 440.

46 An auctioneer sold 45 bags of cotton, each contain-

ing 400 pounds, at 1 mill a pound; what did the whole come to? (R. 61.) Ans. $18.

$$10\not{0} \,\Big|\, \begin{array}{l} 45 \\ 4\not{0} \end{array}$$

47 A mechanic receives $90 for 40 days' work—he worked 12 hours each day; how much was it per hour? (R. 73.) Ans. 18¾cts.

48. A laborer worked 26 days for a farmer, at 87½cts. per day, and took his pay in wheat, at 65cts. per bushel; how many bushels did he receive? Ans. 35.

SECTION IV.

It is an *axiom* in philosophy, that equal causes produce equal effects; and that effects are always proportionate to their causes.

Now, causes and effects that admit of computation, that is, *involve the idea of quantity*, may be represented by numbers, which will have the same relation to each other as the things they represent.

Keeping these premises in view, then, we have a universal rule, applicable to all cases which *can arise* under Proportion, simple or compound, direct or inverse, namely:

Rule.—*As any given cause is to its effect, so is any required cause (of the same kind) to its effect; or, so is another given cause, of the same kind, to its required effect.*

The only difficulty the pupil can experience in this system of proportion, is readily to determine what is cause, and what is effect. But this difficulty is soon overcome, when we consider, that *all action*, of whatsoever nature, must be cause—and whatever is accomplished by that action, or follows such action, must be effect.

EXAMPLES.

1. If 16 horses, in 50 days, consume 128 bushels of

oats, how many bushels will 5 horses consume in 90 days? (W. 113.) Ans. 72.

Here it is evident, that the consumption of oats spoken of, in both the supposition and demand, are the true effects; and the action of the horses, multiplied by the days, must express the amount of cause. We shall therefore state it thus:

Cause.		*Effect.*		*Cause.*		*Effect.*
16	:	128	: :	5	:	[ ]
50				90		

We write the factors, *one* under *another*, as above; their multiplication is understood, but rarely *or never actually* accomplished. The second effect is an unknown term, or answer, required—a bracket, or blank, or point, is left to represent it. When found, the four terms above would be a *perfect Geometrical proportion*, and the product of the extremes equal to the product of the means. In this example, the product of the means is *perfect;* which product, divided by the factors in the extremes, will give what is wanting in the extremes, namly—the answer.

Therefore, agreeably to one mode of performing multiplication and division, we draw a line thus, and cancel down:

16	128
50	5
	90

If $480, in 30 months, produce $84 interest, what capital, in 15 months, will produce $21?

Now capital will not produce interest without time; and, whatever be the rate per cent., the same capital in a double time, will produce a double interest. Therefore,

Cause.		*Effect.*		*Cause.*		*Effect.*
480	:	84	: :	[ ]	:	21
30				15		

Here *one element* of the second cause is wanting; that is, the answer to the question.

In this case, the extremes *are complete;* we will, therefore, divide the product of the extremes by the factors in the mean, and the quotient will give the *definite factor*, or answer, namely—$240.

3 If 7 men, in 12 days, dig a ditch 60 feet long, 8 feet wide, and 6 feet deep, in how many days can 21 men dig a ditch 80 feet long, 3 feet wide, and 8 feet deep?

It is almost too plain for comment, that 7 men, multiplied by 12 days, must be the first cause, and the contents of the ditch they dig, the effect. Therefore,

Cause.		*Effect.*		*Cause.*		*Effect.*
7	:	60	: :	21	:	80
12		8		[ ]		3
		6				8

Here, as in the preceding example, one of the elements of the second cause is wanting; or, rather say a *factor* in the means of a *perfect* proportion, and can be found as above. Ans. $2\frac{2}{3}$ days.

5 If 6 men build a wall in 12 days, how long will it require 20 men to build it? Ans. $3\frac{3}{5}$ days.

Questions of this kind are usually classed under the *single rule of three inverse;* they do in fact, however, belong to compound proportion: but, as one of the terms is the same in the supposition as in the demand, it may be omitted. The term, in the present example, is, *one wall.* If we make the number different in the *two branches* of the question, or connect any conditions with it, such as lengths, breadths, &c., it at once falls under compound proportion of *necessity*, and may be stated thus:

Cause.		*Effect.*		*Cause.*		*Effect.*
6	:	1	: :	20	·	1
12				[ ]		

5 If 4 men, in $2\frac{1}{2}$ days, mow $6\frac{2}{3}$ acres of grass by

working $8\frac{1}{4}$ hours a day, how many acres will 15 men mow in $3\frac{3}{4}$ days, by working 9 hours a day?

Ans. $40\frac{10}{11}$ acres.

Cause.		*Effect.*		*Cause.*		*Effect.*
4	:	$6\frac{2}{3}$	::	15	:	[ ]
$2\frac{1}{2}$				$3\frac{3}{4}$		
$8\frac{1}{4}$				9		

Let the pupil observe, that when a correct statement is made, there will be the same number of *elements*, or *factors*, under the same letters, as in the above example under each *Cause*. We have *men*, *days*, and *hours*, to be multiplied together. When there are fractions in any of the terms, their *denominators* are to be placed over the line from where the *term belongs;* mixed numbers being *previously* reduced to improper fractions.

6 If 12 oz. of wool make $1\frac{1}{2}$ yards of cloth, $\frac{7}{8}$ of a yard wide, how many yards, $1\frac{1}{4}$ wide, will 16 pounds of wool make? Ans $22\frac{2}{5}$ yards.

With this example, some might hesitate as to arranging it under cause and effect, as the *actors*, those who made the cloth, whether many or few, have nothing to do with the question. But we take the *phrase* of the example and say, The wool *makes* the cloth.

Cause.		*Effect.*		*Cause.*		*Effect.*
12	:	$1\frac{1}{2}$	::	16 } oz.	:	[ ]
		$\frac{7}{8}$		16 }		$1\frac{1}{4}$

7 If the transportation of $12\frac{4}{7}$cwt. 206 miles cost \$25.75, how far, at the same rate, may 3 tons and 3qrs. be carried for \$243? Ans. $402\frac{2}{7}$ miles.

In this example, it is indifferent which we take for the cause and which for the effect. We may say, that the money, \$25.75, is the cause of having the weight carried; or, we may say, that carrying the weight is the cause of purchasing the money. There are many questions where it is indifferent which we take for cause, and which for effect. The above example may be stated thus :

Cause.		*Effect.*		*Cause.*		*Effect.*
$25\frac{3}{4}$	:	$12\frac{4}{7}$cwt.	: :	243	:	$60\frac{3}{4}$cwt.
		206 miles.				[ ]

Or thus :

Cause.		*Effect.*		*Cause.*		*Effect*
$12\frac{4}{7}$	:	$25\frac{3}{4}$	: :	$60\frac{3}{4}$	:	243
206				[ ]		

Or thus :

Cause.		*Cause.*		*Effect.*		*Effect.*
$12\frac{4}{7}$	:	$60\frac{3}{4}$	: :	$25\frac{3}{4}$	:	243
206		[ ]				

Or thus :

Effect.		*Effect.*		*Cause.*		*Cause.*
243	:	$25\frac{3}{4}$	: :	$60\frac{3}{4}$	:	$12\frac{4}{7}$
				[ ]		206

These changes show, conclusively, that this method of statement is strictly scientific and philosophical ; and, in all these different arrangements of the terms, the *same terms are multiplied together.*

The most that can be said for the common modes of statements, in the *Double Rule of Three*, is, that the *products, when the terms are multiplied out*, are proportional. But the first and second terms, taken as a whole, express no particular idea or thing ; whereas, in this mode of statement, *the thing*—the philosophical idea—is the *only* sure guide. Nor is this all ; it is very extensive and easy in its application ; it will cover every case that can arise under interest—to find time, rate per cent. &c., and thus do away, or suspend, five or six *special rules*, which encumber every arithmetic. We give a few examp'es to apply this rule.

8 What is the interest of $240 for $3\frac{1}{2}$ years at 6 per cent.?

Cause.		*Effect.*		*Cause.*		*Effect.*
100	:	6	: :	240	:	[ ]
1				$3\frac{1}{2}$		

To obtain the answer from this statement, we perceive that we must multiply the *means* together—i. e. 240×6, the rate, and that by the $3\frac{1}{2}$ years, the time—and divide by 100; *and this is the common rule.*

9 The interest of a certain sum of money, at 6 per cent., for 15 months, was $60; what was the sum?
Ans. $800.

Cause.		*Effect.*		*Cause.*		*Effect.*
100	:	6	: :	[ ]	:	60
12				15		

10 If $800, in 15 months, should gain $60, what would be the rate per cent.? Ans. 6.

Cause.		*Effect.*		*Cause.*		*Effect.*
800	:	60	: :	100	:	[ ]
15				12		

11 Eight hundred dollars was put out at interest, at 6 per cent., and the interest received was $60; how long was it out? Ans. 15 months.

Cause.		*Effect.*		*Cause.*		*Effect.*
100	:	6	: :	800	:	60
12				[ ]		

12 If 12 men, working 9 hours a day for $15\frac{5}{9}$ days, were able to execute $\frac{2}{3}$ of a job, *how many men may be withdrawn* and the residue be finished in 15 days more, if the laborers are employed only 7 hours a day? (W. 109.) Ans. 4 men.

13 The amount of a note, on interest for 2 years and 6 months, at 6 per cent., is $690; required the principal. (R. 172.) Ans. 600.

14 What is the interest of $1248 for 16 days? (R 162.) Ans. $3.28.

Cause.		*Effect.*		*Cause.*		*Effect.*
100	:	6	: :	1248	:	[ ]
12 } days.				16		
30 }						

This can be cancelled down and made very brief.

15 What is the interest of $1200, for 15 days, at 6 per cent.? (R. 162.) Ans. $3.00.

16 How many men will reap 417.6 acres in 12 days, if 5 men reap 52.2 acres in 6 days? (T. 156.) Ans. 20.

17 If a cellar 22.5 feet long, 17.3 feet wide, and 10.25 feet deep, be dug in 2.5 days, by 6 men working 12.3 hours a day, how many days, of 8.2 hours, should 9 men take to dig another 45 feet long, 44.6 wide, and 12.3 deep? (T. 156.) Ans. 12.

18 What is the interest of 160 dollars for 36 days, at 7 per cent.? Stated by cause and effect. Ans. $1.12.

Cause.		*Effect.*		*Cause.*		*Effect.*
100	:	7	: :	160	:	[ ]
12				1.2		

19 The interest of a certain sum, for 36 days, is $1.12—the rate per cent. is 7; what is the sum? Ans. $160.

20 The interest of $160 for 36 days, is $1.12; what was the rate per cent.? Ans. 7.

21 The interest of $160, at 7 per cent., was $1.12; what was the time? Ans. 36 days.

22 In what time will any sum double itself at 6 per cent.? At any per cent.?

Ans. At 6 per cent., in $16\frac{2}{3}$ years; at any per cent., divide 100 by the per cent.

23 If $2\frac{1}{2}$ yards of cloth, $1\frac{2}{5}$ wide, cost 3.37\frac{1}{2}$, how much will $36\frac{1}{2}$ yards cost, $1\frac{1}{2}$ yards wide? Ans. $52.79.

24 If 4 men spend $\frac{3}{4}$ of $\frac{8}{9}$ of $\frac{9}{7}$ of $\frac{14}{15}$ of £30, in $\frac{7}{11}$ of $\frac{9}{13}$ of $\frac{26}{27}$ of $\frac{11}{7}$ of 9 days, how many dollars, at 6 shillings each, will 21 men spend in $\frac{3}{4}$ of $\frac{14}{15}$ of $\frac{6}{7}$ of $\frac{4}{12}$ of 45 days? (Burnham, 142.) Ans. $630.

25 I lend a friend $200 for 6 months; how long ought

he to lend me $1000, to requite the favor—allowing 30 days to a month? Ans. 36 days.

26 If 1000 men, besieged in a town, with provisions for 5 weeks, allowing each man 16 ounces per day, be reinforced with 500 men more—and supposing that they cannot be relieved until the end of eight weeks—how many ounces a day must each man have, that the provisions may last them that time? (W. 183.)
Ans. $6\frac{2}{3}$ ounces

The advocates of this system are of opinion, that there are far too many rules in our common arithmetics; and to reduce them, and thereby simplify the science, they recommend that all the problems, generally arranged under Profit and Loss, Equation of Payments, &c., *should be solved by proportion, and arranged under that head.* In this light, they are more simple and intelligible than they can be made by any *special rules.*

EXERCISES FOR PRACTICE.

1 If I buy cotton-cloth at 2s. per yard, and sell it at 2s. 8d., what do I gain per cent? (W. 134.)
Ans. $33\frac{1}{3}$.

Statement.—If 24 pence gain 8 pence, what will 100 pence gain? Or, 24 : 8 : : 100 : Ans.
Or, 3 : 1 : : 100 : Ans.

2 A merchant bought broadcloth, at $5.50 per yard, and sold it for $6.60; what did he gain per cent? (W. 135.) Ans. 20.

3 If I buy Irish linen at 2s. 3d. per yard, how must I sell it to gain 25 per cent? Ans. 2s. 9d. 3h.

Statement.—If 100 pence return 125 pence, how much must 27 pence return?

Or, 100 : 125 : : 27 : Ans.
Or, 4 : 5 : : 27 : Ans.

4 If, by selling cloth at $6.50 per yard, I lose 20 per cent., what was the prime cost of it? (W. 137.)
Ans. $$8.12\frac{1}{2}$.

That is, if 80 I now receive originally cost me 100, what did 6.50 originally cost?

5 By selling calico at $37\frac{1}{2}$ cents a yard, 50 per cent. was gained; what was the first cost? (R. 185.)

Ans. 25cts.

$$150 : 100 :: 37\tfrac{1}{2} : \text{Ans.}$$
$$\text{Or, } 3 : 2 :: 37\tfrac{1}{2} : \text{Ans.}$$

6 Sold wine at $1.36 per gallon and lost 15 per cent.; what per cent. would have been gained had the wine been sold for $1.856 per gallon? (R. 187.) Ans. 16.

7 Bought 126 gallons of wine for 150 dollars, and retailed it at 20cts. per pint; what was the gain per cent.? (T. 129.) Ans. $34\frac{2}{5}$.

8 If $126.50 are paid for 11cwt. 1qr. 25lbs. of sugar, how must it be sold a pound to make 30 per cent. profit? (W. 135.) Ans. $12\frac{1}{2}$cts.

9 If I buy $12\frac{1}{2}$cwts. of sugar for $140, at how much must I sell it per pound to make 25 per cent.?

Ans. $12\frac{1}{2}$cts.

10 If a firkin of butter, containing 56lbs. cost $7, at how much must it be sold per pound to yield 30 per cent. profit? Ans. $16\frac{1}{4}$cts.

11 What is the whole loss, and what is the loss per cent., in laying out $70 for hats, at $1.75 each, and selling them for 25cts. a-piece less than cost? (Burnham, 173.) Ans. Whole loss $10; loss per 100, $14\frac{2}{7}$.

12 A merchant purchases 180 casks of raisins, at 16 shillings per cask, and sells the same at 28 shillings *per cwt.*, and gains 25 per cent.; what is the weight of each cask? (B. 174.) Ans. 80lb.

	180
	16
4	5
28	112
180	

We multiply 180 by 16, and to add $\frac{1}{4}$ for 25 per cent., we multiply by 5 and divide by 4. Then divide by 28, and it gives cwt.; multiply by 112, and we have pounds; then divide by 180, and we have pounds in each cask. That is, arrange the numbers as above, and *cancel down.*

For other examples, the student is referred to the body of the work.

SECTION V.

Square and Cube Roots.

To work the square and cube roots with ease and facility, the pupil must be familiar with the following properties of numbers.

I. A square number, multiplied by a square number, the product will be a square number.

II. A square number, divided by a square number, the quotient is a square.

III. A cube number, multiplied by a cube, the product is a cube.

IV. A cube number, divided by a cube, the quotient will be a cube.

If the square root of a number is a composite number, the square itself *may be divided into integer square factors*; but if the root is a *prime number*, the square cannot be separated into square factors *without fractions.*

N. B. Substitute the word *cube*, for *square*, in the preceding sentence, and the same remarks apply to cube numbers.

No person can extract roots with any tolerable degree of skill, without being able to recognize the *squares and cubes* of the nine digets as soon as seen.

Numbers,	1	2	3	4	5	6	7	8	9	10
Squares,	1	4	9	16	25	36	49	64	81	100
Cubes,	1	8	27	64	125	216	343	512	729	1000

We here wish to remind the reader, that the pupil is supposed to understand the extraction of the roots in the common way, and we request them not to forget that this is *merely an appendix.*

EXERCISES FOR PRACTICE.

1 What is the square root of 625 ? (R. 220.)
Ans. 25.

If the root is an *integer number*, we may know, by the inspection of the above table, that it must be 25, as the greatest square in 6 is 2, and 5 is the only figure whose square is 5 in its unit place.

Again, take 625
Multiply by . . . . . 4 4 being a square.

2500

The square root of this product is obviously 50 ; but this must be divided by 2, the square root of 4, which gives 25, the root.

2 What is the square root of 6561 ? (R. 220.)
Ans. 81.

As the *unit figure*, in this example, is 1, and in the line of squares in the above table, we find 1 only at 1 and 81, we will, therefore, divide 6561 by 81, and we find the quotient 81 ; 81 is, therefore, the square root.

3 What is the square root of 106729? (T. 170.)
Ans. 327.

As the unit figure, in this example, is 9, if the number is a square, it must divide by either 9, or 49. After dividing by 9 we have 11881 for the other factor, a prime number, therefore its root is a prime number=109. 109, multiplied by 3, the root of 9, gives 327 for the answer.

4 What is the root of 451584 ? (T. 179.)
Ans. 672.

As the unit figure is 4, and in the line of squares we find 4 only at 4 and 64, the above number, *if a square*, must divide by 4, or 64, or by both.

We will divide it by 4, and we have the factors 4 and 112896. This last factor *closes in* 6; therefore, by looking at the table, we see it must divide by 16, or 36, &c. &c.

We divide by 36, and we have the factors 36 and 3136; divide this last by 16, and we have 16 and 196; divide this last fraction by 4, and we have 4 and 49.

Take now our divisors, and last factor, 49, and we have for the original number the product of 4×36×16 ×4×49; the roots of which are 2×6×4×2×7, the products of which are 672, the answer.

5 Extract the square root of 2025. (E. 163.)
Ans. 45.

Divide by 25, and we have its square factors, 25 and 81. Roots of these factors are 5×9=45, the answer.

Again, multiply by the square number 4, when a number ends in 25, and we have 8100, root 90, half of which, because we multiplied by 4, the square of 2, is 45, the answer.

6 What is the square root of 390625? (R. 220.)
Ans. 625.

	390625
Multiply by 4, . . .	4
	1562500
Multiply by 4 again,	4
	6250000

As the number, independent of the ciphers, still ends in 25, we multiply again by 4, and we have 25000000. The root of this is, obviously, 5000. Divide by 2 *three times*, or by 8, and we have 625, the answer.

So far, some may think this *more curious than useful*. However this may be, there are problems where much labor may be saved by attending to the foregoing principles. The following are some of them:

Find a mean proportional between 4 and 256.
Ans. 32.

Find a mean proportional between 4 and 196.
Ans. 28.

Find a mean proportional between 25 and 81.
Ans. 45.

As the above are square numbers, multiply their square roots together for the answers.

EXAMPLES.

1 If 484 trees be planted at an equal distance from each other, so as to form a square, how many will be in a row each way. (T. 171.) Ans. 22.

Factors 4 and 121 2×11 roots=22.

2 A section of land, in the Western states, is a square, consisting of 640 acres; what is the length in rods of one of its sides? (W. 147.) Ans. 320.

Nine out of ten of our teachers *would actually* reduce the acres to square rods, by multiplying by 160, and extract the square root of the product—but this would show too little attention to numbers. Remove one of the ciphers from one number to the other, and we have 64 to be multiplied by 1600, both square numbers, whose roots are 8 and 40—product 320, the answer.

3 What must be the side of a square field, that shall contain an area equal to another field of rectangular shape, the two adjacent sides of which are 18 by 72 rods. (W. 147.) Ans. 36 rods.

18 by 72 is the same as the half of 18 by the double of 72, or 9 by 144, square numbers, roots $3\times 12=36$, the answer.

4 A has two fields, one containing 10 acres and the other $12\frac{1}{2}$; what will be the length of the side of a field containing as many acres as both of them? (R. 220.)
Ans. 60 rods.

22 ,5$\times$160 is the same as 225$\times$16; roots $15\times 4=60$, the answer.

5 What is the mean proportional between 24 and 96?
Ans. 48.

6 What is the mean proportional between 18 and 32?
Ans. 24.

Problems on the Right-angled Triangle.

1 The top of a castle is 45 yards high, and is surrounded with a ditch 60 yards wide; required the length of a ladder that will reach from the outside of the ditch to the top of the castle. Ans. 75 yards.

This is almost invariably done by squaring 45 and 60, adding them together, and extracting the square root; but so much labor *is never necessary when the numbers have a common divisor*, or when the side sought is expressed by a *composite* number.

Take 45 and 60; both may be divided by 15, and they will be reduced to 3 and 4. Square these, 9+16 =25. The square root of 25 is 5; which, multiplied by 15, gives 75, the answer.

2 Two brothers left their father's house, and went, one 64 miles due west, the other 48 miles due north, and purchased farms; how far are they from each other? (E. 171.) Ans. 80 miles.

Divide by . . . 16)64 48

4 3 16+9=25, 5×16=80.

3 The hypothenuse of a right-angled triangle is 520 feet, the base 312 feet; what is the perpendicular?
Ans. 416.

Divide by . . . . 52)520 312

2)10 6

5 3 25—9=16, root 4.

Multiply by 104

Answer 416

4 Required the height of a May-pole, whose top being broken off, struck the ground at the distance of 15 feet from the foot, and measured 39 feet.
Ans. 75 feet.

5 A hawk, perched on the top of a perpendicular tree, 77 feet high, was brought down by a sportsman, standing off 14 rods, on a level with its base; what distance, in yards, did he shoot? (W. 149.)

Ans. 81.15+yards

If this problem is worked with skill, it will be requisite to extract the root of 10 only.

6 If the diagonal of a rectangular field is 40 rods, and one of the sides 32, what is the other? (W. 150.)

Ans. 24.

Cubes and Cube Root.

Cubes, whose roots are composite numbers, may be divided by cube factors. Cube numbers, whose unit figure is 5, may be multiplied by the *cube number* 8, and that period reduced to ciphers.

1 What is the cube root of 91125? Ans. 45.

Multiply by 8

729000

Now 729 being the cube of 9, the root of 729000 is 90; divide this by 2, the cube root of 8, and we have 45, the answer.

2 The contents of a cubical cellar are 1953.125 cubic feet; what is the length of one of its sides? (R. 225.)

Ans. 12.5 feet.

1953.125

Multiply by 8, . . . 8

15625.000

Multiply by 8 again, 8

125.000

The cube root of this is 50; divide by 4, because we multiplied by 8 twice, and we have 12.5 the answer.

3 The number 195112 is a cube; what is its root?

Ans. 58.

The cube numbers are

8, 27, 64, 125, 216, 343, 512, 729.

Comparing these numbers with 195112, and we observe, that the root, in the place of tens, *cannot be* more than 5, and the root, in the place of units, must be some number which, when cubed, give 2 for its unit figure—and 8 is the only figure possible; the root of the whole is, therefore, 58.

4 The number 912673 is a cube; what is its root? Ans. 97.

Observe, the root of the superior period must be 9, and the root of the unit period must be some number which will give 3 for its unit figure *when cubed*, and 7 is the only figure that will answer.

In this manner, we can speedily and easily obtain the cube roots of all cube numbers containing not more than two periods, or determine whether they are cubes or surds.

The following numbers are cubes; required their roots.

1 What is the cube root of 59319? Ans. 39.

2 What is the cube root of 79507? Ans. 43.

3 What is the cube root of 117649? Ans. 49.

4 What is the cube root of 110592? Ans. 48.

5 What is the cube root of 357911? Ans. 71.

6 What is the cube root of 389017? Ans. 73.

8 What is the cube root of 571787? Ans. 83.

When a cube has more than two periods, it can generally be reduced to two by dividing by some one or more of the cube numbers, unless the root is a *prime* number.

The number 4741632, is a cube; required its root. Here we observe, that the unit figure is 2; the unit figure of the root must, therefore, be the root of 512, as that is the only cube of the 9 digits whose unit figure is 2. The cube root of 512 is 8; therefore, 8 is the unit figure in the root, and the root is an *even* number, and can be divided by 2—and, of course, the cube itself can be divided by 8, the cube of 2.

8)4741632
―――――
592704

Now, as the first number was a cube, and being divided by a cube, the number 592704 must be a cube, and, by inspection, as previously explained, its root must be 84, which, multiplied by 2, gives 168, the root required.

The number 13312053, is a cube; what is its root?
Ans. 237.

As there are three periods, there must be three figures, units, tens, and hundreds, in the root; the hundreds must be 2, the units must be 7. Let us then divide the 2d figure, or the tens, *in the usual way*, and we have 237 for the root.

Again, divide 13312053 by 27, and we have 493039 for another factor. The root of this last number must be 79, which, multiplied by 3, the cube root of 27, gives 237, as before.

The number 18609625 is a cube; what is its root?

As this cube ends with 5, we will multiply it by 8:

18609625
8
―――――
148877000

As the first is a cube, this product must be a cube; and, as far as labor is concerned, it is the same as reduced to two periods, and the root, we perceive at once, must be 530, which, divided by 2, gives 265 for the root required.

N. B. If a number, whose unit figure is 5, be multiplied by 8, and *does not result in three ciphers* on the right, the number is not a cube.

To find the Approximate Cube Root of Surds.

The usual way of direct extraction, is too tedious to

be much practiced, if any shorter method can possibly be obtained. By the invention of logarithms, a *very short* method has been found; but, before that event, several eminent mathematicians bestowed much time and labor to obtain short practical rules—and some of their rules are too ingenious and useful to be lost, notwithstanding the invention of logarithms has nearly superceded their absolute value in practice.

There is no exact and *constant relation* between powers and their roots; for this reason, all rules (save by logarithms, and the direct and tedious one,) must be more or less approximate; but, nevertheless, with common judgment and care, we can arrive at results as near as by *the direct* method.

In order to obtain a rule, let us take two cube numbers, as near in value to each other as practicable, and compare them with their roots.

216000 and 226981 are cubes; their roots are 60 and 61.

Now 216000 is not to 226981 as 60 to 61. But let us double the first and add it to the second, and double the second and add it to the first, and we shall have 658981 and 669962, which are to each other *very nearly* as 60 to 61.

Or, by the principles of proportion, the first is to the difference between the first and second, as is the third to the difference between the third and fourth. That is, 658981 : 10981 : : 60 : 1 *very nearly*. Now one is the difference between the two roots, and if the last root, or 61, was unknown, this proportion would give it *very nearly*.

EXAMPLES.

1. Required the cube root of 66.

The cube root of 64 is 4. Now it is manifest, that the cube root of 66 is a little more than 4, and by taking a similar proportion to the preceding, we have

```
64×2=128    2×66=132
      66          64
      ---         ---
     194   :     196 : : 4 : to root of 66.

Or,    194  :  2  :  :  4  :  to a correction.

          194)8.0000(0.04124
               1 76
               ----
                240
                194
                ---
                 460
                 388
                 ---
                  720
```

Therefore, the cube root of 66 is 4.04124

2 Required the cube root of 123.

Suppose it 5; cube it, and we have 125.

Now we perceive, that the cube of 5 being greater than 123, the *correction* for 5 must be *subtracted*

```
     2×125=250    246
Add. . . . . . 123    125
               ---    ---
As . . . . . . 373 : 371 : : 5 : root of 123.

Or,   373  :  2  : :  5  :  correction for 5.

     373)10.0000(0.02681
           7 46
           ----
           2 540              From 5.00000
           2 238              take 0.02681
           -----                   -------
             3020             Ans. 4.97319
             2984
             ----
              360
```

From what precedes, we may draw the following

Rule.—*Take the nearest rational cube to the given number, and call it the assumed cube; or, assume a root to the given number and cube it. Double the assumed cube and add the number to it; also, double the number and add the assumed cube to it. Take the difference of these sums, then say, As double of the assumed cube, added to the number, is to this difference, so is the assumed root to a correction.*

This correction, added to or subtracted from the assumed root, *as the case may require,* will give the cube root very nearly.

By repeating the operation with the root last found as an assumed root, we may obtain results to any degree of exactness; one operation, however, is generally sufficient.

3 What is the cube root of 28? Ans. 3.03658+.

4 What is the cube root of 26? Ans. 2.96249+.

5 What is the cube root of 214?

Ans. 5.98142+.

6 What is the cube root of 346?

Ans. 7.02034+.

The above being very near integral cubes—that is, 28 and 26 are both near the cube number 27, 214 is near 216, &c., all numbers very near cube numbers are *easy of solution.*

We now give other examples, more distant from integral cubes, to show that the labor must be more lengthy and tedious, though the operation is the same.

EXAMPLES.

1 What is the cube root of 3214? Ans. 14.75758.

Suppose the root is 15—its cube is 3375, which, being greater than 3214, shows that 15 is too great; the correction will, therefore, be subtractive.

By the rule, 9964 : 161 : : 15 : 0.2423, the correction.

Assumed root, ... 15.0000
Less 2423

Root nearly 14.7577

Now assume 14.7 for the root, and go over the operation again, and you will have the true root to 8 or 10 places of decimals.

2 What is the cube root of 14760213677 ? Ans. 2453.

Suppose the root 2400, &c. Take the correction to the *nearest unit*, and you will find it 53.

3 What is the cube root of 980922617856 ? Ans. 9936.

Suppose the root to be 10000.

4 What is the cube root of 9 ? Ans. 2.08008.

5 What is the cube root of $9\frac{1}{6}$? Ans. 2.092+

6 What is the cube root of 41 ? Ans. 3.4482+.

When it is requisite to multiply several numbers together and extract the cube root, try to change them into *cube factors*, and extract the root *before* the multiplication.

EXAMPLES.

1 What is the side of a cubical mound equal to one 288 feet long, 216 feet broad, and 48 feet high ? (R. 225.)

The common way of doing this, is to multiply these numbers together and extract the root, a lengthy operation. But, observe, that 216 is a cube number, and $288 = 2\times12\times12$, and $48 = 4\times12$; therefore, the whole product is $216\times8\times12\times12\times12$. Now the cube root of 216 is 6, of 8 is 2, and of 12^3 is 12, and the product of $6\times2\times12 = 144$, the answer.

2 Required the cube root of the product of 448×392, in a brief manner. Ans. 56.

3 If you have a pile of wood 32 feet long, 4 feet

wide, and 4 feet high, what would be the side of a cubic pile containing the same quantity? Ans. 8 feet.

Proposition—The solid contents of cubes or spheres are to each other as the cubes of their like dimensions. (See Geometry.)

EXAMPLES.

1 Mercury is about 2000 miles in diameter, and the earth about 8000; what is their relative magnitudes? Ans. As 1 to 64.

2 Mars is about 4000 miles in diameter, the earth 8000; what is their relative magnitudes? Ans. As 1 to 8.

In the preceding examples *we*, of course, do not cube the numbers given, *but their smallest integral relations.*

3 The diameter of the earth, to that of the sun, is nearly as 1 to $111\frac{1}{2}$; what is their relative magnitudes, or bulks? Ans. As 1 to 1384472 nearly.

4 If a ball, 6 inches in diameter, weigh 32lbs., what will be the weight of a ball, of the same metal, whose diameter is 3 inches? (R. 225) Ans. 4lbs.

6 : 3 : : 2 : 1
8 : 1 : : 32 : 4

5 Suppose an iron ball, of 4 inches in diameter, to weigh 9 pounds; required the weight of a spherical shell of 9 inches in diameter, and 1 inch thick. Ans. 54lbs. 4oz.

6 If a cable, 12 inches about, require an anchor of 18cwt., of what weight must an anchor be for a 15 inch cable? (Pike, 211.) Ans 35cwt. 15lbs.

SECTION VI.

Mensuration, Gauging, &c.

N. B. EVERY problem that follows, can be done with much less labor than they are usually done.

EXERCISES FOR PRACTICE

1 What is the difference of area between a square of 40 rods on a side, and an equilateral rhombus of 40 rods to a side, but of a perpendicular altitude of only 34 rods? (W. 175.) Ans. 240 rods.

2 There is a barn, 50 feet by 36, and 20 feet high to the eaves; how many boards will it take to cover the body, if the boards were all 15 inches wide and 10 feet long? Ans. 275+ boards.

3 On a base of 120 rods in length, a surveyor wished to lay off a rectangular lot of land, to contain 60 acres; what distance in rods must he run out from his base line? (W. 175.) Ans. 80 rods.

4 How many square yards in a triangle, whose base is 48 feet, and perpendicular height $25\frac{1}{4}$ feet? (W. 175.) Ans. $67\frac{1}{3}$ yards.

5 A man bought a farm 198 rods long, and 150 rods wide, and agreed to give $32 per acre; what did the farm come to? Ans. $5940.

N. B. Make no attempt to compute the number of acres definitely.

6. If the forward wheels of a coach are 4 feet, and the hind ones 5 feet in diameter, how many more times will the former revolve than the latter in going a mile? (W. 176.) Ans. 84.

N. B. In this problem use 7 to 22.

7 How many square feet in a board, 2 feet wide at the larger, 1 foot 8 inches at the smaller end, and 14 feet long? (W. 176.) Ans. $25\frac{2}{3}$ feet.

8 The plate supporting the rafters of a house, being 40 feet long, 14 inches wide, and 8 inches thick, how many solid feet does it contain? (W. 177.) Ans. $31\frac{1}{9}$ feet.

12	40	inches
12	12	
12	14	
	8	

Cancel down, and this is the form for all solids.

9 If a pile of wood be 60 feet long, 12 feet high, and 6 feet wide, how many cords does it contain?

Ans $33\frac{3}{4}$ cords.

10 The bin of a granary is 10 feet long, 5 feet wide, and 4 feet high; allowing the cubical contents of a dry gallon to be $268\frac{4}{5}$ cubic inches, how many bushels of grain will it contain? Ans. $160\frac{5}{7}$.

11 If you wanted a bin to contain twice as much as mentioned in the last problem, with a length of 12 feet, and a breadth of 6 feet, of what height must it be? (W. 177.) Ans. $5\frac{5}{9}$ feet.

12 A canal contractor engaged to excavate 2 miles of canal across a plane, at 8cts per cubic yard—the canal to be 54 feet wide at top, 40 at bottom, and $4\frac{1}{2}$ feet deep; what did it amount to? Ans. $6617.60.

13 There is a circular cistern, of uniform diameter, whose depth is 8 feet, and diameter 5 feet; what is its capacity, allowing 231 inches to the gallon, and how much would its capacity be increased by adding 6 inches to its diameter?

Ans. { 1175.04 gallons its capacity,
$246\frac{3}{4}$ gallons increase.

14 What would be the produce of a kernel of wheat in 11 years, at 20 fold, the produce of each year being sowed the next—allowing 5000 kernels to a quart? (W. 166.) Ans. 64000000 bushels.

N. B. If we *blindly* perform all the labor indicated by set rules, the above would be a tedious operation; but it is *extremely* brief in the hands of a skillful operator.

15 The length of a room being 20 feet, its breadth 14 feet 6 inches, and its height 10 feet 4 inches; how much will the coloring come to at 27cts. per square yard, deducting a fire-place of 4 feet by 4 feet 4 inches, and two windows, each 6 feet by 3 feet 2 inches? (R. 232.)

Ans. $19.73.

16 What will the paving of a foot-path come to, at 18 cents per square yard, the length being 35 feet 4 inches, and the breath 8 feet 3 inches? (R. 233.) Ans. $5.83.

17 What will it cost to roof a building 40 feet long, the rafters on each side being 18 feet 6 inches long, at $3.50 per 100 square feet? (R. 233.) Ans. $51.80.

18 There is a block of marble, in the form of a parallelopiped, whose length is 3 feet 2 inches, breadth 2 feet 8 inches, and depth 2 feet 6 inches; what will it cost at 81cts. per cubic foot? (R. 234.) Ans. $17.10.

19 What will it cost to build a wall 320 feet long, 6 feet high, and 15 inches thick, allowing 20 bricks to the solid foot, at 5\frac{875}{1000}$ per thousand bricks? (R. 234.) Ans. $282.

20 How many bricks 8 inches long, 4 inches wide, and 2$\frac{1}{4}$ inches thick, will it take to build a wall 120 feet long, 8 feet high, and 1 foot 6 inches thick? (R. 234.) Ans. 34560.

21 What will it cost to build a brick wall 240 feet long, 6 feet high, and 3 feet thick, at $3.25 per 1000 bricks—each brick being 9 inches long, 4 inches wide, and 2 inches thick? (R. 234.) Ans. $336.96.

22 A ship's hold is 75$\frac{1}{2}$ feet long, 18$\frac{1}{2}$ wide, and 7$\frac{1}{4}$ deep; how many bales of goods 3$\frac{1}{2}$ feet long, 2$\frac{1}{4}$ deep, and 2$\frac{3}{4}$ wide, may be stowed therein, leaving a gangway, the whole length, of 3$\frac{1}{4}$ feet wide? (Pike, 470.)

Ans. 385.4+.

N. B. Do this by one operation—after taking out the gangway *mentally*, by subtraction.

23. A stick of timber is 16 inches broad and 8 inches thick; how many feet in length must be taken to make 20 solid feet? Ans. 22$\frac{1}{2}$.

24 There is a square pyramid, each side of whose base is 30 inches, and whose perpendicular height is 120 inches, to be divided by sections, parallel to its base, into three equal parts; required the perpendicular height of each part. (P. 371.)

Ans. The height of the lower section is 15.2 inches; the height of the middle section is 21.6 inches; the height of the top section is 83.2 inches.

N. B. In solving this problem, remember that solids,

of the same shape, are to each other as the cubes of their *like sides.*

25 A man wishes to make a cistern of 8 feet in diameter, to contain 60 barrels, at 32 gallons each, and 231 cubic inches to a gallon; what shall be the depth of the cistern?

$60 \times 32 \times 231$ gives the cubic inches the cistern is to contain.

This divided by the circular end, expressed in inches, will give the depth in inches.

Now, $\frac{8 \times 12 \times 22}{7}$ = the circumference in inches.

But the circumference of any circle, multiplied by $\frac{1}{4}$ of its diameter, giver its area.

Then, ... $\frac{8 \times 12 \times 22 \times 24}{7}$ = the area.

Hence, ... 8̸	6̸0̸ 5
1̸2̸	3̸2̸ 4̸
2 2̸2̸	2̸3̸1̸ 2̸1̸ 7
2 8̸ 2̸4̸	7

$7 \times 7 \times 5 = 245$; which, divided by 4, gives $61\frac{1}{4}$ *inches for the answer.*

SECTION VII.

Miscellaneous Examples.

1 One-half, one-third, and one-fourth of a certain number, added together, make 130; what is the number?

Ans. 120.

To solve this by arithmetic, we must consider the number as the whole of a thing, or a unit. Then $\frac{1}{2}+\frac{1}{3}+\frac{1}{4}=\frac{6}{12}+\frac{4}{12}+\frac{3}{12}$ or $\frac{13}{12}$. Then, by proportion, if $\frac{13}{12}$ make 130, what will 1, or $\frac{12}{12}$, make? That is,

$$\frac{13}{12} : 130 :: \frac{12}{12}$$

By multiplying the first and last terms by 12, the proportion becomes

13 : 130 : : 12

Or, ... 1 : 10 : : 12

2 One-fourth of a certain number exceeds one-sixth of the same number by 20; what is the number?

Here, again, the number must be considered as a single *thing;* then $\frac{1}{4}$ of it is not, strictly, one-fourth of a unit, but $\frac{1}{4}$ of that number, or that thing. In algebra, this thing, or number, would be represented by some letter, as x or y—and the question is strictly an algebraical one. But all such questions in algebra, can be solved by fractions and proportion in arithmetic; and, indeed, all questions, that involve simple equations only, can be resolved by arithmetic—but by algebra they are much easier.

In days that *were*, we always found a *rule in arithmetic called Position*, which included such problems as the preceding, and some few others, but could not take in any questions involving powers, or roots—as powers and roots are not in arithmetical proportion to each other.

For example, 16 and 64 are square numbers, and their square roots are 4 and 8, or as 1 to 2; but the numbers themselves, 16 and 64, are to each other as 1 to 4, a different relation from the roots.

Example: A man, having a purse of money, being asked how much was in it, answered, The square root of it, added to the half of it, made 220 dollars; how much was in the purse?

It is evident, that this question must be excluded from any proportional operation; for, unless we first suppose the *right number*, the result of the supposition will not be to the given result as the supposed number to the true number—and when this *proportion* fails, supposition, that is, "*Position fails;*" and if we suppose the true number, it is then an operation of chance, and is, in fact, no problem at all.

True it is, we can give a *rule*, or rather give *orders*, which, followed, will reduce the problem, *but it will not be arithmetic;* and, to put into an arithmetical work what is not arithmetic, we hold to be deceptive and improper.

We have remarked, that the solution of these problems are much easier by algebra than by arithmetic; but we would remind the pupil, that the solution of a problem is a *small object* compared with a *principle* involved in a solution, or with a knowledge of the science of numbers. As one step to secure this latter object, we give a few more such problems.

3 A post is $\frac{1}{5}$ in the earth, $\frac{3}{7}$ in the water, and 13 feet above the water; what is the length of the post?

Ans. 35 feet.

Add $\frac{1}{5}$ and $\frac{3}{7}$; not $\frac{1}{5}$ and $\frac{3}{7}$ of the number 1, but $\frac{1}{5}$ and $\frac{3}{7}$ of the *whole post.* These parts, added together, make $\frac{22}{35}$; the remaining $\frac{13}{35}$ must be 13 feet. Then, by proportion,

$$\tfrac{13}{35} : 13 :: \tfrac{35}{35}$$

Or, ... $13 : 13 :: 35 : 35$ the answer.

4 A and B have the same income. A contracts an annual debt amounting to $\frac{1}{7}$ of it, B lives upon $\frac{4}{5}$ of it; at the end of ten years, B lends to A money enough to pay off his debts, and has 160 dollars to spare; what is the income of each?

Ans. \$280.

If B lives on $\frac{4}{5}$, he saves $\frac{1}{5}$; out of this he pays A's debts, $\frac{1}{7}$. Hence, from $\frac{1}{5}$ subtract $\frac{1}{7}$, and there remains $\frac{2}{35}$. This, in 10 years, is worth 160 dollars; therefore, in 1 year it is worth 16 dollars. Now, by proportion,

$\frac{2}{35} : 16 :: \frac{35}{35} :$ the answer;

Or, . . $2 : 16 :: 35 :$ the answer;

Or, . . $1 : 8 :: 35 : 280$, the answer.

5 Of the trees in an orchard, $\frac{3}{4}$ are apple trees, $\frac{1}{10}$ pear trees, and the remainder peach trees—which are 20 more than $\frac{1}{8}$ of the whole; what is the whole number in the orchard?

Ans. 800.

The apple, the pear, and the peach trees, make the whole; therefore, add $\frac{3}{4}+\frac{1}{10}+\frac{1}{8}=\frac{60}{80}+\frac{8}{80}+\frac{10}{80}=\frac{78}{80}$.

This wants $\frac{2}{80}$, or $\frac{1}{40}$, of being the whole; therefore, we *must conclude*, that the 20, not taken into the account, is $\frac{1}{40}$ of the whole. Hence, $20\times40=800$; or, by proportion,

$\frac{1}{40}$: 20 : : $\frac{40}{40}$: answer.

That is, 1 : 20 : : 40 : 800, the answer.

6 A, B, and C, would divide $200 between them, so that B may have $6 more than A, and C $8 more than B; how many dollars for each? (R. 229.)

Ans. A 60, B 66, C 74.

Observe, that A has the least sum, B $6 more than A, and C $14 more than A; hence, $20 is to be reserved, and the remaining 180 to be divided *equally among them;* which, of course, gives $60 to A, and $60+6 to B, and 60+6+8 to C.

7 A gentleman bought a chaise, horse, and harness for $378—the horse came to twice the price of the harness, and the chaise to twice the price of both the horse and harness; what did he give for each? (R. 229.)

This problem is generally given under position, but it is really one in simple division. Take the idea of shares. Divide the money into shares: it will take 1 share to purchase the harness, 2 shares to purchase the horse, 6 shares to purchase the chaise; therefore, divide the whole into 9 shares, and we shall have $42 for one share, i. e. $42 for the harness, $84 for the horse, $252 for the chaise.

In conclusion, we would caution the young arithmetician against imbibing the idea, that he understands arithmetic, from works on arithmetic alone. We must rise above a plane, to have a fair view of the objects on its surface, and we must rise above arithmetic before we can understand all its scientific relations.

Nor must we conclude that we are perfect in numbers, because we may excel our comrades in disentangling knotty questions. Science is a different thing from acuteness at solving intricacies; and all men, of true science, more or less despise all things *intended to puzzle*, for there are enough objects of useful inquiry and investigation, on which to expend all our powers of mind.

Algebra is but a continuation of arithmetic; and as soon as we acquire a good practical knowledge of arithmetic, so as to have a clear comprehension of fractions and general proportion, we may, *yea, we should*, commence algebra. But when we advance in algebra, we should be careful not to look down with the spirit of contempt on arithmetic—we may study the one in order to understand the other. The following properties of numbers, the student must take as facts, *unless* he is an algebraist. They cannot be demonstrated without the aid of that science.*

1 *If from any number, the sum of its digits be subtracted, the remainder is divisible by* 9.

Take, for example, 34; subtract 7, and we have 27, which is divisible by 9. Again, take 438; subtract the sum of 4+3+8=15, and we have 423=9×47, and so with *any other* number.

2 *If the sum of the digits of any number be divisible by* 9, *the number itself is divisible by* 9.

Thus the numbers 72, 81, 99, 171, 387, 51489, &c., the sum of whose digits is divisible by 9, are themselves divisible by 9.

All numbers divisible by 9, are, of course, divisible by 3.

3 *If the sum of the digits of any number be divisible by* 3, *then the number itself is divisible by* 3.

Thus 18, 27, 54, 75, 111, 123, 258, &c. &c., are all divisible by 3.

4 *If from any number the sum of the digits standing in the* ODD *places be subtracted, and the sum of the digits standing in the* EVEN *places be added, then the result is divisible by* 11.

Take any number, say 785432; then subtract the sum of 2+4+8=14, and add 3+5+7=15, or, in this example, *add* 1, and we have 785433=11×71403.

---

* The demonstrations of these properties may be found in Bridge's Algebra, section XLII.

5 *If the sum of the digits standing in the* EVEN *places, be equal to the sum of the digits standing in the* ODD *places, then the number is divisible by* 11.

Thus the numbers 121, 363, 12133, 48422, &c., are all divisible by 11.

Our numbers increase in a ten-fold proportion, from the right to the left, which is called, *the root of the scale;* but, if the scale was 7, in lieu of ten, then, what is *now* true for 9 and 11, would be *then* true for 6 and 8.

6 *We may change numbers from the scale of ten to any other scale, by dividing the number by the number denoting the scale, continually saving the remainders and forming a new number by them.*

Example—Change 63 into an equivalent number, wherein the value of the digit shall increase in a five-fold proportion; in other words, where the *root of the scale* shall be 5.

5)63
5)12(3= first remainder.
5)2(2= second remainder.
0(2= third remainder.

Now 223 is the value of 63, in a scale where the numbers increase in a five-fold proportion. Change 3714 to its equivalent value on a scale of 4. Ans. 322002.

Numbers may also be considered as arising from the *continued multiplication of certain factors.*

A perfect number, is one which is equal to the sum of all its divisors. They are not numerous, and may be expressed thus:

$2\ (2^2-1)=2\times 3=6.$

$2^2(2^3-1)=4\times 7=28.$

$2^4(2^5-1)=16\times 31=496.$

$2^6(2^7-1)=64\times 127=8128.$

$2^{10}(2^{11}-1)=1024\times 2047=2096128.$

A PRACTICAL SYSTEM

OF

BOOK-KEEPING,

FOR

MECHANICS AND RETAILERS.

Book-Keeping is the method of recording a systematic account of business transactions.

It is of two kinds—*Single* and *Double Entry*. The former, only, will be noticed in this work. On account of the simplicity of Single Entry, it is, perhaps, the best which can be recommended to farmers, mechanics, and retailers. It consists of two principal books—the *Day Book*, or *Waste Book*, and the *Leger*, and one auxiliary book, the *Cash-Book*.

THE DAY BOOK.

This book is ruled with a column on the left hand for the date, and three columns on the right, the first, for the folio or page of the Leger, to which the account is transferred; and the last two for dollars and cents.

This book exhibits a minute history of business transactions in the order of time in which they occur, with every circumstance, necessary to render the transaction plain and intelligible.

				$	cts.
	David Judkins, Dr.				
Jan. 4	To 10 lbs. coffee at 17 cts.	$1 70			
	" 25 lbs. sugar, at 10 cts.	2 50	1	4	20
	Timothy W. Coolidge, Dr.				
" 4	To 1 bl. sugar, weighing 135 lbs. neat, at 8½ cts.	$11 47½			
	" 1 bag coffee, 98 lbs. at 15 cents,	14 70	1	26	17½
	Geo. H. Eaton, Dr.				
" 6	To 1 bl. flour,	$3 87½			
	" 1 lb. Y. H. Tea,	1 12½			
	" 1 keg lard, neat weight 60 lbs., at 6¼ cts.	3 75	1	8	75
	David Judkins, Cr.				
" 8	By cash on account,		1	3	00
	James Wilson, Dr.				
" 11	To 3 weeks boarding, at 2 dollars per week,		2	6	00
	Hiram Ames, Dr.				
" 11	To 12 yards broadcloth at 6 dollars per yard,	$72 00			
	" 30 yds. muslin at 14 cts.	4 20	2	76	20
	Cr.				
" 14	By an order on J. Jones, for Groceries,	$51 00			
	" Cash,	20 00	2	71	00
	Timothy W. Coolidge, Cr.				
" 14	By a bill of carpenter work,		1	25	00
	James Wilson, Dr.				
" 15	To 1 bl. vinegar,	$3 25			
	" 1 keg 10d. nails, weight				
	" 111 lbs. at 8 cts.	8 88	2	12	13

Date	Account		$	cts.
	George Hamilton, Cr.			
Jan. 19	By 1 set of Fancy chairs,	3	25	00
	George H. Eaton, Cr.			
" 21	By cash, to balance account,	1	8	75
	Robert Young, Dr.			
" 25	To cash on account, $3 00			
	" 10 lbs. N. O. sugar, at			
	" 10 cents, 1 00			
	" 12 lbs. coffee at 16⅔ cts. 2 00			
	" ½ lb. Y. H. Tea, 62½	3	6	62½
	Jackson Moore, Dr.			
" 29	To 21 lbs. Ham, at 10 cts. $2 10			
	" 1 box soap, 30 lbs. at 5			
	cents, 1 50	3	3	60
	David Judkins, Dr.			
" 30	To 12 bls. apples at 75 cents,	1	9	00
	Cr.			
	By 47 bush. corn, at 20 cents,	1	9	40
	James Wilson, Cr.			
" 31	By cash on account,	2	15	00
	Thomas Hilton, Dr.			
Feb. 3	To cash, $5 00			
	" 1 lb. Y. H. Tea, 1 06			
	" 16 lbs. Rice, at 6¼ cts. 1 00	2	7	06
	Cr.			
	By 13 days labor, at 87½ cts.	2	11	37½
	George Hamilton, Dr.			
" 10	To 1 canister Imperial Tea, 14			
	lbs. at $1 87½ per lb.	3	26	25
	Hiram Ames, Cr.			
" 13	By order on E. Disney for goods,	2	10	00

Date	Account			$	cts.
	Robert Young, Dr.				
Feb. 15	To 1 box sperm candles,				
	" 25 lbs., at 30 cts. per lb.	$7 50			
	" 2 bu. dried fruit, at $1				
	" 25 per bushel,	2 50	3	10	00
	James Wilson, Dr.				
" 21	To 10 gals. molasses, at 40 cents,	$4 00			
	" 4 lbs. Old Hyson Tea, at 93 cents,	3 72	2	7	72
	Robert Young, Cr.				
" 27	By 12 cords of wood at	$2 25	3	27	00
	Ames & Smith, Dr.				
March 6	To 18 lbs. sole leather, at				
	" 25 cents,	$4 50			
	" 1 side upper leather,	2 75			
	" 3 calf skins, $1 25,	3 75	4	11	00
	Hiram Ames, Dr.				
" 6	To 500 feet white pine boards, at				
	" $12 50 per M.	$6 25			
	" 25 bu. potatoes, at 50 cts.	12 50			
	" 1 ton of Hay,	10 00	2	28	75
	David Judkins, Dr.				
" 8	To 200 lbs. flour, at	$1 75	1	3	50
	Thomas Hilton, Dr.				
" 8	To cash paid his order to William Coolidge,		2	18	75
	George Hamilton, Dr.				
" 15	To 1 copy Whelpley's Compend,	$1 25			
	" 1 ream letter-paper,	4 50			
	" 1 doz. Spelling books,	1 00	3	6	75

Date		Account		$	cts.
		Theodore P. Letton, Dr.			
Mar.	20	To sharpening his plough, $1,00 " Shoeing his horse, 1,62½ " Repairing chain, 25	4	2	87½
		Timothy W. Coolidge, Dr.			
"	"	To 2 qrs. tuition of himself at evening school, at $3 per qr.	1	6	00
		Thomas Hilton, Cr.			
"	23	By the hire of his horse 10 days, at 62½ cts. per day,	2	6	25
		Ames and Smith, Cr.			
"	25	By 1 hhd. sugar, weight 1317 lb. neat, at 7½ cts.	4	98	77½
		T. P. Letton, Dr.			
"	26	To I ream wrapping paper, $1,62½ " 1 beaver hat, 5,00 " 1 set silver tea-spoons, 6,00	4	12	62½
		Henry C. Sanxay, Cr.			
"	28	By my order on him in favor of Jno. Torrence for stationary,	4	12	87½
		Ames and Smith, Dr.			
"	30	To 2000 ft. clear pine boards, at $20 per M. $40,00 " 500 common do. at $8, 4,00 " 5000 shingles, at $2,25 11,25 " Cash to balance account, 32,52	4	87	77
"	31	Jackson Moore, Cr.			
		By painting my house,	3	21	00

			$	cts.
	Thomas Hilton, Dr.			
March 31	To 4 bu. wheat, at $1 25, $5 00			
	" 1 bl. mess pork, 9 00			
	" 2 bu. salt, at 50 cts. 1 00			
	" 8 lbs. brown sugar, 11 cts. 88	2	15	88
	George Hamilton, Dr.			
" 31	To 12 cedar posts, at 25c. $3 00			
	" 1 plough, 9 37½			
	" 1 scythe, 1 62½	4	14	00
	Jackson Moore, Dr.			
" 31	To repairing his wagon and plough,	3	6	00
	Thomas Hilton, Cr.			
" 31	By 1 pair shoes, $1 50			
	" 1 mahogany table, 12 50	2	14	00
	George Hamilton, Cr.			
" 31	By an order on J. Hulse, $5 00			
	" cash, 6 50	3	11	50

END OF THE DAY BOOK.

THE LEGER.

This book is used to collect the scattered accounts of the Day Book, and to arrange all that relates to each individual, into one separate statement. The business of transferring the accounts from the Day Book to the Leger, is called *posting.*

The Leger is ruled with a double line in the middle of the page, to separate the debits from the credits. Each side has two columns for dollars and cents, one for the page of the Day Book, from which the particular item is brought, and a column for the date.

When an account is posted, the page of the Leger on which this account is kept, is written in the column for that purpose in the Day Book, and also the page of the Day Book from which the account was posted, is written in the 2d column of the Leger.

In posting, begin with the first account in the Day Book, which you will perceive is the name of David Judkins. Enter his name in the first page of the Leger, in a large, fair hand, with Dr. on the left and Cr. on the right.—As there are several articles charged to D. Judkins on the 4th of January, instead of specifying each article in the Leger, we merely say, *For Sundries,* and enter the amount in the proper columns—see Leger, page 1.

The Leger has an index or alphabet, in which the names of persons are arranged under their initial letters, with the page in the Leger where the account may be found.

ALPHABET TO THE LEGER.

A		I J		R	
Ames, Hiram -	2	Judkins, David -	1		
Ames & Smith -	4				
B		K		S	
Balance - -	5			Sanxay, H. C. -	4
C		L		T	
Coolidge, T. W. -	1	Letton, T. P. -	4		
D		M		U	
		Moore, Jackson -	3		
E		N		V	
Eaton, George H.	1				
F		O		W	
				Wilson, Jas. -	2
G		P		X	
H		Q		Y	
Hilton, Thomas -	2			Young, Robt. -	3
Hamilton, Geo. -	3			Z	

1 *Dr.* *David Judkins,* *Cr.*

1833.									
Jan. 4	To sundries	2	4	20	Jan. 8	By cash	2	3	00
" 30	" Apples	3	9	00	" 30	" Corn	3	9	40
Mar. 7	" Flour	4	3	50		" *Bal.*		4	30
			16	70				16	70
Apl. 1	To balance of account bro't down,		4	30					

NOTE.—The Dr. on the left hand side of the page signifies *debtor*, and that the sums entered on that side of the page, are those for articles sold to others, and for which *they owe you.* The Cr. on the right hand side signifies *credit*, or that the sums entered on that side of the page are for articles received of the person under whose account they stand, and for which *you owe him.*

Timothy W. Coolidge.

Jan. 4	To sundries	2	26	17½	Jan 14	By work	2	25	00
Mar. 20	" Tuition	5	6	00	Apl. 4	" *Balance*		7	17½
			32	17½				32	17½
Apl. 1	To balance		7	17½					

NOTE.—The Dr. side of this account shows the amount of articles Mr. Coolidge has received of me, and the Cr. side shows what I have received of him. It appears that the total amount of my account against him is $32 17½, from which deduct the $25 00 (which stands to his credit on the right of the account) and there will remain a balance of $7 17½ due me.

George H. Eaton.

Jan. 6	To sundries	2	8	75	Jan 21	By cash	3	8	75

NOTE.—This account presents equal sums on both sides; hence it is evident that I owe J. H Eaton nothing, and that he owes me nothing. The account is fully closed.

Dr.			*Hiram Ames.*				*Cr.*		2
Jan11	To sund.	2	76	20	Jan. 11	By sund.	2	71	00
Mar.6	" Corn	4	28	75	Feb. 13	" Order	3	10	00
					Apl. 1	" *Bal.*		23	95
			104	95				104	95
Apl. 1	To *Bal.*		23	05					

Thomas Hilton.

Feb. 3	To sund.	3	7	06	Feb. 3	By labor,	4	11	37½
Mar 8	" Cash.	4	18	75	Mar23	" hire of horse,	5	6	25
" 31	" do.	6	15	88	" 31	" Sund.	6	14	00
					Apl. 1	" *Bal.*		10	06½
			41	69				41	69
Apl. 1	To *Bal.*		10	06½					

James Wilson.

Jan. 11	To boarding,	2	6	00	Jan. 31	By cash	3	15	00
" 15	" Sunds.	2	12	13	April 1	" *Bal.*		10	85
Feb. 21	" do.	4	7	72					
			25	85				25	85
April 1	To balance bro't down,		10	85					

NOTE.—When an account is *settled* only, and not *fully paid*, as in the above, and several preceding accounts, the *balance*, whether it be in your favor or against you, is brought down and placed distinctly by itself, and serves for the beginning of a *new* account, as you perceive has been done in the above example, the balance being $10 85.

Dr. *George Hamilton.* **Cr.**

Feb. 10	To Tea,	3	26	25	Jan. 19	By chairs,	3	25	00
Mar. 15	" Sund.	4	6	75	Mar 31	" Sunds.	6	11	50
" 31	" do.	6	14	00	Apl. 1	*Balance,*		10	50
			47	00				47	00
April 1	To *Bal.*		10	50					

Robert Young.

Jan. 27	To sundries,	3	6	62½	Feb. 27	By wood,	4	27	00
Feb. 15	" do.	4	10	00					
Apl. 1	" *Bal.*		10	37½					
			27	00				27	00
					Apl. 1	By *Bal.*		10	37½

NOTE.—In the above account the difference between the Dr. and Cr. side is $10 37½, by which I perceive that the balance *against me*, in favor of Robert Young, is $10 37½.

Jackson Moore.

Jan. 29	To Sundries,	3	3	60	Mar. 31	By painting,	5	21	00
Mar. 31	" Sunds.	6	6	00					
Apl. 1	" *Bal.*		11	40					
			21	00				21	00
					Apl. 1	By balance bro't down,		11	40

Dr.		Ames & Smith.							Cr.		
Mar. 6	To	Sundries	4	11	00	Mar.25	By sugar,	5	98	77
" 30	"	do.	5	87	77					
				98	77				98	77

NOTE.—This account, like the one on the first page, is fully closed, the amount on the Dr. side being just equal to that on the Cr.

T. P. Letton.

Mar.20	To Sundries	5	2	87½	Ap. 1	*By Bal.*		15	50
" 26	" do.	5	12	62½					
			15	50				15	50
April 1	To balance		15	50					

NOTE.—In the above account there is no sum on the Cr side, and the inference is, that T. P. Letton owes me $15 50 for sundry articles expressed in detail, in the Day Book, page 5.

Henry C. Sanxay.

Apl 1	*To balance,*		12	87½	Mar.28	By my order,	5	12	87½
			12	87½				12	87½
					Apl. 1	*By bal.*		12	87½

NOTE.—In this account it will be perceived that as there is no amount charged to H. C. Sanxay, on the Dr. side, I owe him $12 87½.

5	*Dr.*				*Balance.*				*Cr.*
April 1	To D. Judkins,	1	4	30	April 1	By R. Young,	3	10	37½
	" T. W. Coolidge,	1	7	17½		" J. Moore,	3	11	40
	" H. Ames,	2	23	95		" H. Sanxay,	4	12	87½
	" T. Hilton,	2	10	06½					
	" J. Wilson,	2	10	85					
	" G. Hamilton,	3	10	50					
	" T. P. Letton,	4	15	50					
			82	34				34	65

NOTE.—This account exhibits the exact state of your books. It is made from the preceding accounts in the Leger. The Dr. side is an exhibit of the amounts due *to you* by others, and the Cr. side the amounts due *by you* to others. It is not strictly necessary that this account should be introduced in the Leger in single entry: it will be found convenient, however, to balance the book at stated intervals, and transfer the *balances* to the new accounts below, as in the preceding Leger, and when that is done, a *balance account* like the above, will be found convenient, as presenting, at *one view*, the exact state of your Leger.

FORM OF A BILL FROM THE PRECEDING.

Mr. David Judkins,

To Edward Thomson, Dr.

January 4	To	10 lbs. coffee at 17 cts.	$1 70		
		25 lbs. sugar at 10 cts.	2 50		
				4	20
" 30		12 bbls. apples at 75 cts.		9	00
March 8		200 lbs. flour at $1 75,		3	50
				16	70
		Cr.			
January 8	By	Cash,	$3 00		
30		47 bushels corn at 20 cts.	9 40		
				12	40
		Errors excepted. Balance due,		4	30
		Rec'd payment in full, EDWARD THOMSON.			

THE CASH BOOK.

The Cash Book is used to record the daily receipts and payments of money. It is ruled nearly the same as the Leger; the Dr. side exhibits the amount of money *received*, and the Cr. side, the amount *paid out*. Subtract the sum of the Cr. from that of the Dr. and the balance will always be equal to the amount of cash on hand.

FORM OF A CASH BOOK.

Dr.				*Cash.*			*Cr.*
1833				1833			
Jan. 1	To cash on hand,	73	81	Jan. 1	By rent of house paid T. P. Letton,	18	00
" 1	Cash rec'd of J. Young,	16	40		Paid note to R. Hand,	50	00
" 1	" " H. Sanxay,	25	00	" 1	Family expenses,	4	37
" 1	" " D. Judkins.	10	00	" 1	*By cash on hand,*	52	84
		125	21			125	21
Jan. 2	To cash on hand,	52	84		By cash paid Ames &		
" 2	" of T. Coolidge,	23	16	Jan. 2	Smith,	20	00
" 2	Cash found on Main St.	29	00	" 2	*By cash on hand,*	85	00
		105	00			105	00
Jan. 3	To cash on hand,	85	00				

www.ingramcontent.com/pod-product-compliance
Lightning Source LLC
LaVergne TN
LVHW010252110826
845151LV00004B/1451

* 9 7 8 1 4 2 5 5 2 2 7 0 4 *